GEO142 Lab Manual

Geology of Pacific Northwest
Volcanoes, Mountains, and Earthquakes

Written and compiled by
Reanna Camp-Witmer, Shannon Othus-Gault,
and Autumn Christensen

with additional contributions from Michelle Harris and Mariah Tilman

Geology 142 Lab Manual

ISBN: 978-1-943536-60-3
Edition 2 Fall 2019

Chemeketa Press

Chemeketa Press is a nonprofit publishing endeavor at Chemeketa Community College that works with faculty, staff, and students to create affordable and effective alternatives to commercial textbooks. All proceeds from the sales of this textbook go toward the development of new textbooks. To learn more, visit www.chemeketapress.org.

Publisher: David Hallett
Director: Steve Richardson
Managing Editor: Brian Mosher
Instructional Editor: Stephanie Lenox
Design Editor: Ronald Cox IV
Cover Design: Brandi Harbison
Interior Design and Layout: Brice Spreadbury, Cierra Maher
Cover Photo: High-Angle Photography of Green Trees, by Spencer Watson is in the public domain (https://unsplash.com/photos/dEOC8M_lmxl).

Acknowledgments

Text and image acknowledgments appear on pages 105-106 and constitute an extension of the copyright page.

Printed in the United States of America.

Contents

Lab 1
What a Geologist Sees

Name _____   Date _____

Purpose: An introduction to volcanology and the scientific method.
Materials: Images in the manual or powerpoint images presented by your instructor.
Instructions: Answer the following questions.

Figure 1. Belknap Crater area.

1. Look at figure 1. What do you think you are looking at here?

2. Describe the process that you believe deposited these rocks.

Figure 2. Crooked River rhyolite and Newberry basalt.

3. Look at figure 2. Describe the two different rock types shown. How are they different?

4. Which of the two rocks shown is younger? Why?

Figure 3a. Crater Lake.

Figure 3b. Wizard Island, Crater Lake.

5.   Look at figure 3a. How do you think this lake formed?

6.   Look at figure 3b. What is the apparent difference between the island rock and the rocks found in the crater wall?

7.   Look at 3b again. Which feature formed first, the crater or the island?

Figure 4. Yellowstone panorama.

8. Describe the landscape in figure 4. Why do you think there is a boardwalk for patrons of the park?

9. Why are no plants growing near the water? Support your hypothesis.

10. Describe the color of the water. How does it look different from most lakes and ponds you see?

Figure 5. The Spires, Crater Lake.

11. Describe the appearance of the spires in figure 5. How does the color and shape of the of them change vertically?

12. These spires are on the flank of Crater Lake. What do you think created them?

Lab 2
Plate Tectonics

Name _____   Date _____

Purpose: During this lab you will be looking at various data that has been collected by scientists to better understand the theory of Plate Tectonics. Be sure to read each section's instructions and answer all questions.

Background: Plate Tectonics is a relatively new theory, considering that the science of geology is over 200 years old and Plate Tectonics was established about 60 years ago. Naturalist Alfred Wegener built a base of knowledge of what would become the theory of Plate Tectonics in 1915 but referred to his theory as Continental Drift. He surmised that based on the shape of the continents as well as similarities of rocks and fossils in continents separated by oceans, that the continents, were all connected at one point. However, the scientific community initially scoffed at the theory, especially since Wegener could show no proof of what forces compelled the plates of the Earth to move.

Materials:

- Seismology Map
- Volcanology Map
- Ruler
- Calculator
- Access to the Internet

Instructions: Look at the figures and answer the related questions.

PART 1: MAGNETIC ANOMALIES

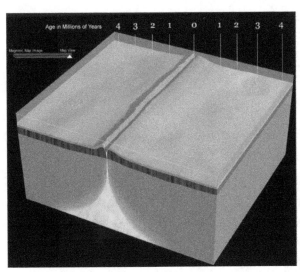

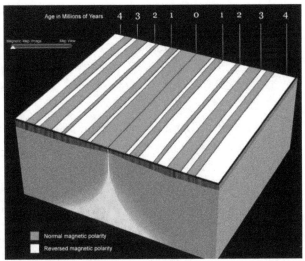

Figure 1. Shape of the seafloor, including the mid-ocean ridge.

Figure 2. Magnetic polarity of the ocean floor over the past 4 million years.

1. Based on your knowledge of seafloor spreading, how many magnetic reversals have been recorded in this picture (do not include the current magnetic polarity in your answer)?

2. Approximately how many years ago did the last period of reverse polarity end? In what direction would a compass needle have pointed during this time?

3. Could you expect to find similar magnetic polarity stripes on iron-rich rocks on land? If you could, would those stripes be as useful as the ocean floor in determining past orientations of continents? Explain your answers.

PART 2: EARTHQUAKES, VOLCANOES, AND PLATE TECTONICS

For this section please use the volcano and earthquake maps provided (figures 6 and 7, pages 13 and 14). Carefully examine the map keys before you answer the following questions.

1. Match the following landform to the appropriate tectonic plate boundary (one landform will have more than 1 boundary):

_____ Mountain Chain	**a)** Divergent
_____ Volcanic Island Arc	**b)** Ocean-Ocean Convergent
_____ Volcanic Arc	**c)** Continental-Ocean Convergent
_____ Mid-Ocean Ridge	**d)** Continental-Continental Convergent
_____ Trench	
_____ Rift Valley	

2. Notice that Transform plate boundaries were not included as a type of plate boundary in the above list. That is because they are not associated with any large landforms. Why not?

3. Where are the deeper earthquakes (the blue and green dots) located? Give the name of a country or specific region they are associated with. Why do you think these deeper earthquakes are forming in these areas?

4. Generally, in which continents are the majority of volcanoes located? Specifically, where do you see a majority of the world's volcanoes?

5. What is a pattern that you see regarding the placement of volcanoes and the location of earthquakes? (There are many possible answers here)

6. Find a location where volcanoes and earthquakes do not mirror each other. Why do you think this is occurring? What makes this location special?

PART 3: THE SPEED OF THINGS

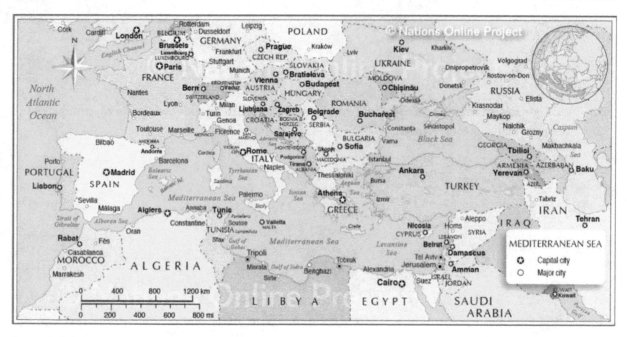

Figure 3. Political Map of the Mediterranean region.

1a. The African Plate is moving at approximately 2.15 cm/yr to the northeast. Using the scale provided, measure the distance between Tunis, Tunisia, and Genoa, Italy.

1b. Using both the rate of plate motion and the distance between these two locations, calculate the number of years it will take for the Mediterranean Sea to close completely. Show your work below.

2. What can we presume will be created when Africa collides with Europe?

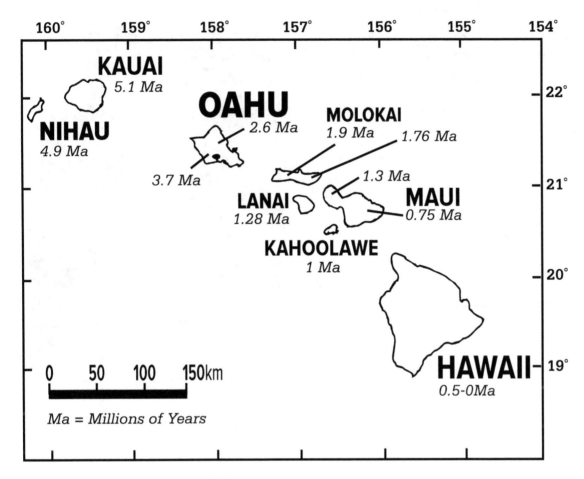

Figure 4. Age of the Hawaiian Island chain.

3a. How far apart are Hawaii and Oahu in kilometers? In centemeters? (Measure from the hot spot, i.e. from the southeast corner of the island of Hawaii).

3b. How fast is the Pacific Plate moving in cm/yr? Show your work.

4. Based on the ages of the Hawaiian Islands, what direction is the Pacific Plate moving?

PART 4: PLATE SPECIFICS

1. Based on the overall southwestern motion of the North American Plate shown in figure 5, what should happen to the Juan de Fuca Plate in the future? What would that mean in terms of volcanoes and earthquakes, in say, 20 million years in our future?

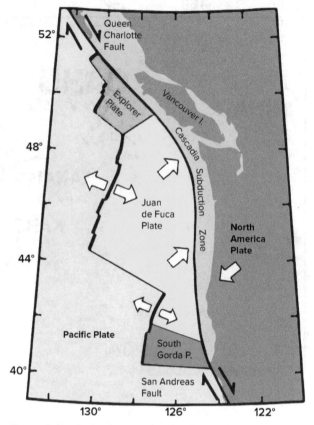

Figure 5. Tectonic plates in the Pacific Northwest.

2. With a partner, choose a plate that you would like to study. Find out what other plates it has boundaries with and the types of behavior seen at these boundaries. Also describe the overall movement of the plate and the various phenomena or landforms created by the movement of your plate. Please make sure to write at least a paragraph.

SCIENTIFIC SPECIALTY: VOLCANOLOGY

Red dots indicate currently or historically active volcanic features

This list obtained from the Smithsonian Institution

This map is part of "Discovering Plate Boundaries," a classroom exercise developed by Dale S. Sawyer at Rice University (dale@rice.edu). Additional information about this exercise can be found at http://terra.rice.edu/plateboundary .

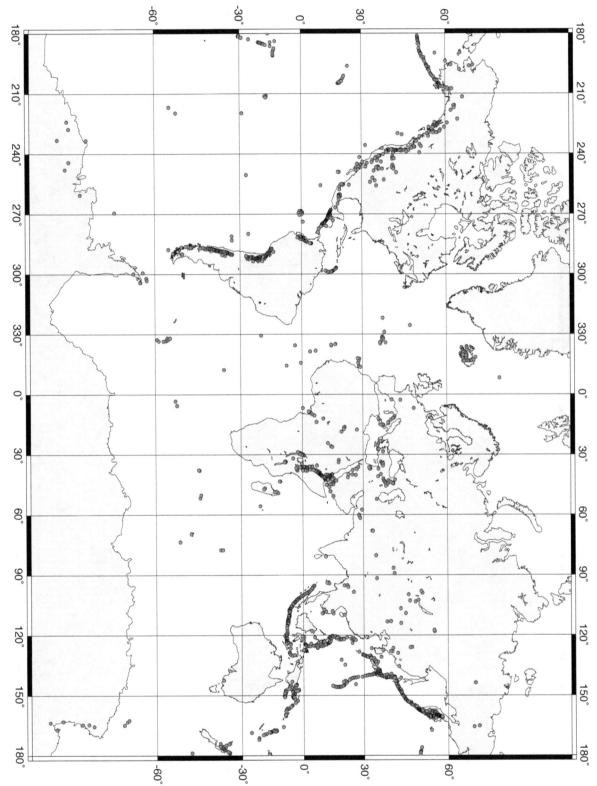

Figure 6. Volcanology map. Courtesy of Dale S. Sawyer and Rice University (http://plateboundary.rice.edu/downloads.html).

SCIENTIFIC SPECIALTY: SEISMOLOGY

Earthquake Locations 1990 - 1996 (Magnitudes 4 and greater)

Color indicates depth: Red 0-33 km, Orange 33-70 km, Green 70-300 km, Blue 300-700 km

This map is part of "Discovering Plate Boundaries," a classroom exercise developed by Dale S. Sawyer at Rice University (dale@rice.edu). Additional information about this exercise can be found at http://terra.rice.edu/plateboundary .

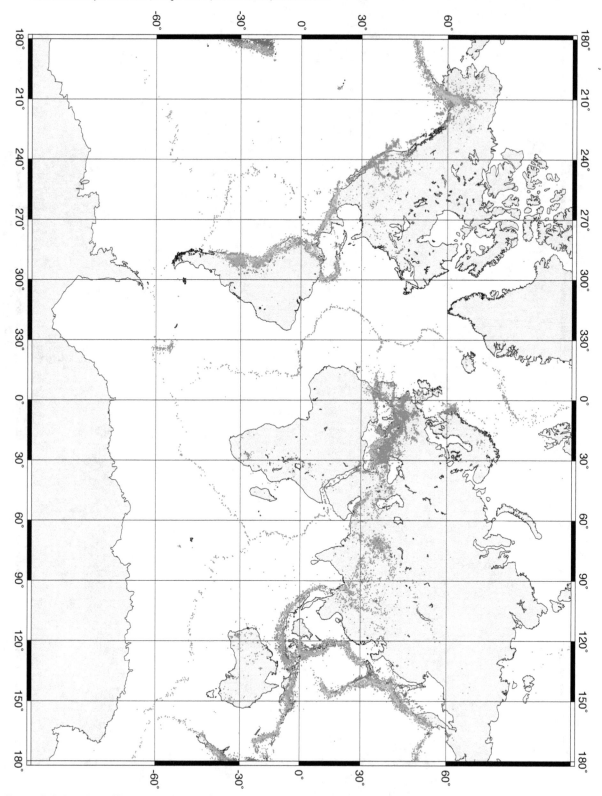

Figure 7. Seismology (Earthquake) Map. Courtesy of Dale S. Sawyer and Rice University (http://plateboundary.rice.edu/downloads.html).

Lab 3
Minerals

Name _____   Date _____

Overview: To explore the nature of minerals and familiarize ourselves with some of the most common minerals in Earth's crust, especially the igneous rocks of Oregon.

Materials:

- Mineral samples
- Glass plate
- Copper penny
- Porcelain streak plate
- Hand lens
- Magnet

- Dilute hydrochloric acid
- Metric balance
- Graduated cylinder
- Beaker or squeeze water bottle
- Calculator

Background: In order to understand the geologic processes that shape the surface of the Earth, you must first have a basic understanding of the nature of the materials that make up Earth's crust. Minerals are the fundamental building blocks of all rocks. When we refer to the "composition" of rocks, we are referring to their mineral makeup, or **mineralogy**. In this and several other lab exercises, you will see how relevant mineral composition is to understanding volcanism and other igneous processes that form the majority of the Pacific Northwest landscape.

Geologically speaking, a **mineral** is a naturally occurring, inorganic solid with an orderly (repetitive) crystalline structure and a definite chemical composition. All minerals can be designated by their chemical formulas and atomic arrangements.

There are over 4,000 named minerals on Earth, and each one has a distinctive set of physical properties such as hardness, cleavage, color, density, luster, fracture, magnetism, etc. These and other physical properties are commonly used for quick identification of minerals since their internal structure and chemical composition are difficult to measure without elaborate equipment. Most physical means of identification require only the skill of the student and some easily obtained everyday materials. It is important to note that totally pure minerals are rarely found in nature. In general, the more pure the mineral specimen, the more recognizable its properties will be.

The first step in learning how to identify minerals is to become acquainted with the various properties that individually or collectively characterize a mineral species. Carefully follow the sections below and you will see how these properties are determined using the five mineral samples.

HARDNESS

Hardness is the resistance of a mineral to abrasion (scratching). Hardness can be determined by scratching a mineral of unknown hardness against a substance of known hardness. The harder substance will scratch or leave a groove in the softer one (i.e., the softer one gets scratched, and the harder one does the scratching).

Hardness is measured on a relative scale using **Mohs Scale of Hardness**, developed by Friederich Mohs in 1822. It consists of ten index minerals arranged in order of increasing hardness where the softest mineral (talc) has a hardness of 1 and the hardest mineral (diamond) has a hardness of 10. As a means of comparison, the Mohs hardness values of some common objects are listed in Table 2.

Mohs Hardness Scale	
1	Talc
2	Gypsum
3	Calcite
4	Fluorite
5	Apatite
6	Orthoclase
7	Quartz
8	Topaz
9	Corundum
10	Diamond

Table 1. Mohs Hardness Scale.

Mohs Hardness for Common Items	
2.5	Fingernail
3.5	Copper penny
4.5	Wire nail
5.5	Glass or knife blade
6.5	Porcelain streak plate

Table 2. Mohs Hardness for common items.

Figure 1. Testing hardness with a glass plate.

First, test a mineral specimen using the glass plate provided (see figure 1). Find an edge or a corner of the specimen that looks the most like the rest of the mineral (i.e., make sure it is the same color and appearance, not an impurity). It is important that you **do not try to scratch the mineral with the glass**, as it could chip or break in your hand. Hold the plate firmly on the table top while trying to scratch it with your specimen. Very hard substances (H>7) produce deep scratches, but substances of H~6 may produce only faint scratches. If the sample does not scratch the glass (H<5.5), try scratching it with your fingernail. If you cannot scratch it with your fingernail, the sample must have a hardness value of 2.5-5.5. On the other hand, if you can scratch it with your fingernail, the hardness of the mineral is <2.5.

Helpful hint: if it appears that a groove has formed, double check by brushing it with your finger. Occasionally, the softer substance will "powder off" on the harder substance and give the appearance of a scratch on the harder one.

LUSTER

Luster refers to the appearance of the mineral's surface in reflected light. All minerals are said to have either a metallic or non-metallic luster. A *metallic* luster has the appearance of polished metal (such as gold, silver, brass, or copper) and is totally opaque to light. Exposed surfaces can tarnish over time and give a dull appearance (or "sub-metallic"), so it is helpful to look at a fresh surface of the mineral. A *non-metallic* luster does not resemble metal and can be any color. There are several types of non-metallic lusters, such as *vitreous* (the luster of glass, any color); *resinous* (the luster of resin or tree sap); *brilliant* (having the appearance of a gem); *pearly* (the luster of a pearl or the inside of a clamshell); or *earthy* (dull or dirt-like appearance).

CLEAVAGE AND FRACTURE

The way a mineral breaks[1] provides a lot of insight because it indicates the internal arrangement of atoms within the mineral and the strength of the bonds holding those atoms together. A mineral will tend to break along the plane of its weakest atomic bonds. **Cleavage** (*cleave* = to break) describes the tendency of a mineral to break consistently along definite smooth planes (flat surfaces) called **cleavage planes**. A mineral may exhibit distinct cleavage along one or more cleavage planes, or it may exhibit no cleavage (see Fracture).

[1]Breaking a mineral is normally accomplished by striking it with a hammer or by dropping it onto a hard surface. It does not mean sawing, shaping, or carving.

Before deciding if a mineral has cleavage, rotate it under a good light source and observe whether there is some position in which the surface of the specimen appears to "light up". Cleavage planes reflect more light than other breakage surfaces and thus appear shinier. Some minerals exhibit "perfect" or "excellent" cleavage with obvious large, shiny, and flat surfaces, sometimes in characteristic shapes[2]. Others exhibit "good" or "poor" cleavage that is less obvious.

Cleavage can be further described by the number of directions cleavage planes and the angles at which they intersect. The number of directions is the number of sets of parallel cleavage planes:

◆ **Basal** cleavage (1 direction) will appear as thin plates or sheets (like the side of a book) when viewed from the "side" (the "top" and "bottom" are parallel, thus it has 1 direction). Minerals in the mica family usually exhibit excellent basal cleavage.

◆ **Prismatic** cleavage (2 directions) is often the most difficult to identify because these minerals rarely break into characteristic shapes. The cleavage planes may intersect at right angles (90°), as in feldspar, or at other angles (e.g., 56° and 124°), as in hornblende. If a mineral with 2 directions of cleavage is a cube, 4 of the sides would be more reflective than the other 2, meaning that the 4 shiny sides are cleavage planes (2 directions) and the other 2 are fracture surfaces.

◆ **Cubic** cleavage (3 directions at 90°) is when cleavage planes meet at 90° (right angles). Often, a mineral with cubic cleavage will easily break into cubes or will have numerous right angle intersections when inspected closely. Common examples are galena and halite.

◆ **Rhombic** cleavage (3 directions not at 90°) is similar to cubic, but the intersections of the planes are not at 90°. A mineral with rhombic cleavage might look like a deck of cards where the top has been slid over relative to the bottom. One common example is calcite.

◆ **Octahedral** cleavage (4 directions) looks like two pyramids that share the same square-shaped base. This occurs more often when a sample is pure. Two examples are fluorite and diamond.

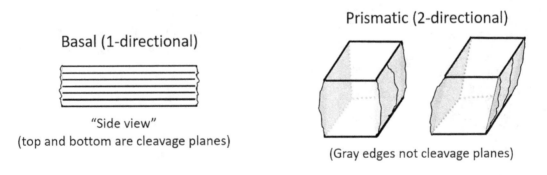

Basal (1-directional)

"Side view"
(top and bottom are cleavage planes)

Prismatic (2-directional)

(Gray edges not cleavage planes)

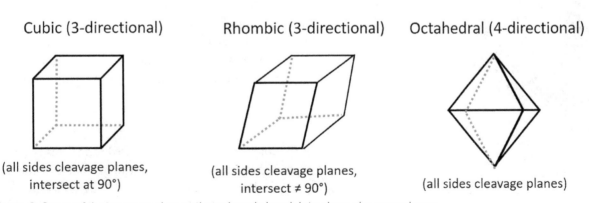

Cubic (3-directional)

(all sides cleavage planes,
intersect at 90°)

Rhombic (3-directional)

(all sides cleavage planes,
intersect ≠ 90°)

Octahedral (4-directional)

(all sides cleavage planes)

Figure 2. Some of the common shapes that minerals break into along cleavage planes.

[2]Characteristic shapes can also be due to crystal habit (see Crystal Form section).

If a mineral does not exhibit cleavage, it is said to have fracture. **Fracture** is typically the result of bonds that are equally or similarly strong in all directions. As a result, breakage surfaces are not flat, but *irregular* (uneven), *conchoidal*[3] (smooth curved surfaces or concentric ridges), *hackly* (jagged edges), or *splintery*. Additionally, fracture surfaces tend to be much duller than cleavage planes.

Although cleavage and fracture describe how a mineral breaks, it is not usually necessary to individually break a mineral specimen because most of the samples you will see have already been broken from larger pieces. Do not break a mineral specimen without your instructor's approval!

SPECIFIC GRAVITY

The **density** of an object is its mass per unit of volume, or in other words, how much it weighs compared to how much space it takes up. It is represented by the following formula where mass is in grams (g) and volume is in cubic centimeters (cm³):

$$\frac{\text{Mass (g)}}{\text{Volume (cm}^3\text{)}}$$

Geologists typically use a related measure, **specific gravity**, to describe the density or "heft" of a mineral or rock. Specific gravity (SG) is a ratio of the density of a substance (such as a mineral) to the density of water (1.0 g/cm3). This cancels out the units so essentially it is just an expression of density without units. Most rock-forming minerals have a specific gravity between 2.3 and 3.5 making this property difficult to compare among most samples. Some metallic minerals, however, are noticeably greater [e.g., galena (PbS) is 7.5 and gold (Au) is 17.2].

In Part 1 of this lab exercise, you will not be calculating the density or specific gravity of all the mineral samples you will be identifying, but rather comparing this quality among mineral specimens. In order to

do this, it is ideal to select two nearly equal-sized samples and decide which is heavier. Pick up the samples provided one at a time, then try holding one in each hand. You will notice that each seems to have a certain "heft," or specific gravity. In Part 2 of this lab exercise, we will be calculating the density and specific gravity of 5 minerals and/or rocks that your instructor has selected.

Figure 3. A few of the many color varieties of quartz.

COLOR

Color is typically the most obvious property, but it should be documented with caution. Although there are minerals that have consistent colors, many are found in numerous colors. For example, cinnabar is always red and malachite is always green, but quartz and fluorite come in a variety of colors. Color variations within the same type of mineral can occur due to the presence of impurities or defects within the mineral structure. When identifying minerals, it is important to keep this in mind.

STREAK

Streak is the color of a mineral in powdered form. This is a useful property because although the outward color of a mineral can vary, its streak will usually not. Streak is determined by rubbing the specimen on a piece of unglazed porcelain (streak

Figure 4. Testing streak with porcelain streak plate.

[3]Conchoidal fracture is also displayed by some rocks with consistent mineral compositions (such as obsidian, or micrite).

plate) causing part of the mineral to rub off onto the streak plate. This will only work for minerals whose hardness is less than that of the streak plate (H=6.5) because if the mineral is harder, it will scratch the streak plate instead. Many glassy minerals have a white streak that is difficult to see on a white streak plate. This may only be visible on close inspection. Minerals with a metallic luster typically have a very distinct and easily identifiable streak.

CRYSTAL FORM (HABIT)

A crystal is a solid that has flat surfaces arranged in a definite geometric shape. The crystal is really the outward expression of the internal atomic arrangement of the mineral. Crystal form is the common or characteristic shape of a mineral. Since all minerals have orderly atomic structures, all minerals can form perfect crystals if their growth is unrestricted. This is rare, however, because most crystals get crowded together as they grow and their shape becomes less characteristic. Additionally, some minerals have two or more forms. For example, the mineral pyrite is commonly found in either cubes or pyritohedrons (having twelve pentagonal faces).

Knowledge of the various crystal forms can be very useful in the identification of a mineral, but it may also be difficult to differentiate between crystal form and cleavage because cleavage planes may be parallel to crystal faces (e.g., calcite).

OTHER PROPERTIES

The following properties will be useful in identifying only certain minerals.

Reaction to Acid: Certain minerals, such as carbonates (with CO_3 in chemical formula) will react with acid by effervescing or "fizzing." When a small amount of dilute acid (typically hydrochloric acid, HCl) is applied to a fresh surface of a carbonate mineral, a chemical reaction occurs and carbon dioxide (CO_2) gas is released as bubbles (figure 5). This is especially obvious in the mineral calcite ($CaCO_3$), which will effervesce vigorously. Two other carbonate minerals are malachite [$(CuOH)_2CO_3$)] and dolomite [$CaMg(CO_3)_2$]. These minerals also effervesce, but the reaction is much slower. To magnify the reaction, place a drop of acid on a freshly powdered surface and view under your hand lens. After you have tested a mineral sample with acid, wipe off the liquid with a paper towel. Do not get acid on your skin. Although it is dilute, it can still burn your skin or damage your clothing. If you get acid on your skin, rinse it off with cool water.

Magnetism: Many minerals contain the element Iron (Fe). If a mineral has a high enough iron concentration, it will attract a magnet. To test this, simply hold a hand magnet close to the sample and see if you feel a pull. The mineral magnetite (Fe_3O_4) will strongly attract a magnet, but hematite (Fe_2O_3) will only weakly attract a magnet.

Double Refraction: Double refraction is an optical property present in very few minerals where an image passing through a transparent piece of the mineral appears to be doubled. It occurs

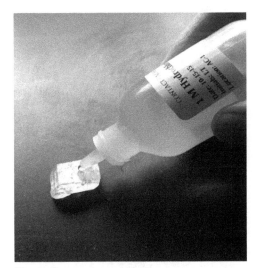

Figure 5.

Figure 6. Calcite showing double refraction.

when a light beam is split into two perpendicular directions due to the arrangement of the atoms. Calcite ($CaCO_3$) is one example, but note that only a transparent piece will show it.

Striations: Striations are fine, parallel grooves that may be present on the crystal face or cleavage plane of some minerals. Quartz crystals and plagioclase feldspars exhibit striations.

Taste: Some minerals have a distinct taste. For example, halite (NaCl) is chemically identical to table salt (and therefore tastes like it!). Sylvite (KCl), which is often used as a salt substitute for those watching their sodium (Na) intake, has a more bitter taste. Kaolinite, a clay mineral, tastes like aluminum. Do not taste a mineral without asking your instructor first.

Feel: Several minerals have a distinct feel. Talc, for example, has a soapy feel (like a bar of soap) while graphite feels greasy.

Odor: Certain minerals give off a distinct smell. This is especially true if a mineral contains the element sulfur (S). Galena (PbS) and sphalerite (ZnS) both smell strongly of sulfur (like rotting eggs, a match head, or sometimes well water). Kaolinite will give off a distinct "earthy" odor if you breathe on it before sniffing.

PART 1: IDENTIFYING IMPORTANT MINERALS

Using the provided mineral samples, determine the properties and name of each. Use mineral tables on pages 23-28 as a guide.

Metallic Luster

H < 5½

Hardness	Color	Streak	Luster	Cleavage/ Fracture	Other Properties	Name and Chemical Composition
1.0–2.0	lead-gray to black	heavy gray	metallic to dull	basal cleavage, easily weathers into fine flakes	low SG (2.1) for mineral with metallic luster; will write on paper; greasy feel	GRAPHITE; C
2.5	bright lead gray	heavy gray	metallic	excellent cubic cleavage	sulfur smell; high SG (7.5)	GALENA; PbS
2.5–3.0	brassy-yellow	yellow-gold	metallic	hackly or irregular fracture	very high SG (17.2)	GOLD; Au
2.5–3.0	silver-steel	silver-white	metallic	hackly or irregular fracture	high SG (10.3)	SILVER; Ag
2.5–3.0	copper (penny)	copper-red	metallic	hackly or irregular fracture	high SG (8.8); often has green discoloration from weathering	COPPER; Cu
3.0	brassy-brown, mottlings of blue or purple	gray-black	metallic	irregular fracture	commonly massive, micro-crystalline	BORNITE; Cu$_5$FeS$_4$
3.5	brassy-yellow	black	metallic	irregular fracture	commonly massive	CHALCOPYRITE; CuFeS$_2$
3.5–4.0	brown to orange to red	yellow	highly vitreous	dodecahedral – 6 directions!	Faint sulfur smell; SG = 4.0	SPHALERITE; ZnS
1.5–5.5	orange, brown, or yellow	yellow	sub-metallic & earthy varieties	irregular fracture	non-crystalline (amorphous)	LIMONITE; Fe$_2$O$_3$nH$_2$O (see also non-metallic)
5.0–6.0	black	black	dull to sub-metallic	conchoidal fracture	non-crystalline (amorphous)	PSILOMELANE; BaMn$_9$O$_{16}$(OH)$_4$

Metallic Luster

H > 5½

Hardness	Color	Streak	Luster	Cleavage/ Fracture	Other Properties	Name and Chemical Composition
5.5-6.0	silver-gray-white	black	metallic	prismatic cleavage	bitter smell on fresh surfaces	ARSENOPYRITE; FeAsS
6.0-6.5	brass-yellow	greenish-black	metallic	irregular fracture	"Fool's Gold"; often forms as cubes or pyritohedrons	PYRITE; FeS₂
5.5-6.5	steel-gray dull black, or deep, earthy red, depending on variety	red	metallic	irregular fracture	three common varieties: specular (bright metallic), oolitic (containing small, spherical masses), sedimentary (earthy and usually deep red). Hardness can be difficult to determine since some specimens easily crumble."	HEMATITE; Fe₂O₃ (see also non-metallic)
6.0	dark gray to black	black	metallic to dull black (if weathered)	irregular fracture	can be strongly magnetic	MAGNETITE; Fe₃O₄

Non Metallic Luster

H < 2½

Hardness	Color	Streak	Luster	Cleavage/ Fracture	Other Properties	Name and Chemical Composition
1.0	white, gray, light yellow, pink, blue or light green	white	pearly, silky	basal cleavage, but weathers easily	very soft; soapy feel	TALC; $Mg_3Si_4O_{10}(OH)_2$
2.0	clear and colorless to frosty white	white	vitreous to dull	basal cleavage	leaves a frosty appearance	GYPSUM; $CaSO_4 \cdot 2H_2O$
2.0	colorless, white, or with shades of red or yellow	white	vitreous	perfect cubic cleavage	salty and bitter taste	SYLVITE; KCl
2.0-2.5	white, gray-brown	white	vitreous, transparent in thin sections	perfect basal cleavage	part of mica group; thin, flexible sheets or "books"	MUSCOVITE (Mica); $KAl_2(AlSi_3O_{10})(OH)_2$
2.0-2.5	dark green	faint green to yellow	pearly or dull	basal cleavage	forms short prisms that split into flexible sheets	CHLORITE; hydrous K-Mg-Al silicate
1.5-2.5	bright yellow	white to yellow	usually earthy and opaque	irregular fracture	strong odor	SULFUR; S
2.0-2.5	white	white	earthy	irregular fracture	easily leaves powder behind on surfaces; earthy-type smell when damp	KAOLINITE; $Al_2Si_2O_5(OH)_4$
1.5-5.5 (variable)	orange, brown, or yellow	yellow, orange, or reddish	sub-metallic & earthy varieties	irregular fracture	non-crystalline (amorphous)	LIMONITE; $Fe_2O_3nH_2O$ (see also metallic)

Non Metallic Luster

H 2½ - 5½ WITH CLEAVAGE

Hardness	Color	Streak	Luster	Cleavage/ Fracture	Other Properties	Name and Chemical Composition
2.5	colorless to white	white	glassy	perfect cubic cleavage	tastes like salt; soluble in water, transparent to translucent	HALITE; NaCl
2.5-3	dark brown to black	slight greenish	glassy or splendent	perfect basal cleavage	part of mica group; thin, flexible sheets or "books"	BIOTITE (Mica); $K(Mg,Fe)_3(AlSi_3O_{10})(OH)_2$
3.5-4.0	usually a shade of pink, but can be white, brown	white	vitreous to pearly	perfect rhombic	effervesces weakly in acid; usually small, elongate crystals with saddle-like faces	DOLOMITE; $CaMg(CO_3)_2$
3.5-4.0	brown to orange to red	yellow	highly vitreous	dodecahedral – 6 directions!	Faint sulfur smell; SG = 4.0	SPHALERITE; ZnS
3.0	usually colorless to white, can be orange, gray, yellow, or brown	white	glassy	excellent rhombic cleavage	strongly reacts to dilute acid	CALCITE; $CaCO_3$
3-3.5	colorless to white (often with tints of brown, yellow, or red)	white	glassy or pearly	excellent cleavage	high SG for non-metallic (4.5); forms short tabular crystals or rose shaped masses	BARITE; $BaSO_4$
4.0	almost any color (colorless, light green, blue, yellow, purple)	white	glassy	octahedral	Exhibits fluorescence, may have slight greasy feel	FLUORITE; CaF_2
5.0-6.0	dark green to black	white	vitreous	two cleavage directions at about 60° & 120°	forms elongated crystals; sometimes bladed; most common amphibole	HORNBLENDE; $CaMg_3(SiO_3)_4$
5.0-6.	dark green to black	white	vitreous	two cleavage directions at about 90°	most common pyroxene	AUGITE; (very complex silicate)

Non Metallic Luster

H 2½ - 5½ ~~WITH~~ *without* CLEAVAGE

Hardness	Color	Streak	Luster	Cleavage/ Fracture	Other Properties	Name and Chemical Composition
3.5–4.0	azure blue	blue	dull to vitreous	irregular fracture	crusts, laminated masses, or prisms; reacts slowly with acid; usually found with malachite	AZURITE; $Cu_3CO_3(OH)_2$
3.5–4.0	bright green (Paris green)	green	dull to vitreous	irregular fracture	crusts, laminated masses, or prisms; reacts slowly with acid; usually found with azurite	MALACHITE; $Cu_2CO_3(OH)_2$
2.5–3.5	white, gray, red, rust color	light	dull, earthy	irregular fracture	contains rounded masses with an earthy matrix	BAUXITE; hydrous oxides of aluminum
5.0	green, yellow, variable	white	vitreous to sub-resinous	conchoidal fracture	hexagonal crystals	APATITE; $Ca_5(PO_4)_3(OH,F,Cl)$
2.5–5.0 (varies)	light to dark green, sometimes with gray	white	waxy, silky	conchoidal/splintery	some fibrous varieties (asbestos)	SERPENTINE; $Mg_6(Si_4O_{10})(OH)_8$
5.5–6.5	steel-gray, dull black, or deep, earthy red, depending on variety	red	metallic	irregular fracture	three common varieties: specular (bright metallic), oolitic (containing small, spherical masses), sedimentary (earthy and usually deep red). Hardness can be difficult to determine since some specimens easily crumble."	HEMATITE; Fe_2O_3 (see also non-metallic)

Non Metallic Luster

H 2½–5½ → H > 5.5

Hardness	Color	Streak	Luster	Cleavage/ Fracture	Other Properties	Name and Chemical Composition
5.0–6.0	dark green to black	white	vitreous	two cleavage directions at about 60° & 120°	forms elongated crystals; sometimes bladed; most common amphibole	HORNBLENDE; $CaMg_3(SiO_3)_4$
5.0–6.0	dark green to black	white	vitreous	two cleavage directions at about 90°	most common pyroxene	AUGITE; (very complex silicate)
6.0	usually varying shades of pink ("salmon" pink) sometimes white	white	vitreous to pearly	two directions nearly at 90°	thin veinlets on cleavage planes; variety of Potassium Feldspar	ORTHOCLASE; $KAlSi_3O_8$
6.0	black to blue-black	white	high vitreous	two directions nearly at 90°	thin veinlets on cleavage planes; will show iridescence	LABRADORITE; (Ca,Na) $(Al,Si)AlSi_2O_8$ (Plagioclase Feldspar)
6.0	white, gray, green	white	vitreous to pearly	two directions nearly at 90°	striations on some cleavage planes	ALBITE; $NaAlSi_3O_8$ (Plagioclase Feldspar)
4.5–6.5	blue to gray-blue	white	vitreous	perfect basal	bladed crystals; hardness varies - 4.5 (long axis), 6.5 (short axis)	KYANITE; Al_2SiO_5
6.5–7.0	light to dark green	white	vitreous	conchoidal fracture	often a mass of small crystals, with an appearance like granulated sugar	OLIVINE; $(MgFe)_2SiO_4$
7.0	colorless, white, purple, pink, yellow, brown	white	vitreous	conchoidal fracture	many varieties (e.g., smoky, milky, citrine, rose, amethyst, tiger's eye, etc.)	QUARTZ; SiO_2
7.0	can be white, gray, blue, brown, red	white	vitreous to sub-vitreous to dull	conchoidal fracture	many varieties (e.g., jasper, flint, chalcedony, agate, etc.), can form very sharp edges	MICROCRYSTALLINE QUARTZ; SiO_2
7.0–7.5	deep red to reddish-brown		vitreous to resinous	sub-conchoidal fracture	usually occurs as perfect crystals	GARNET; complex silicate
7.5	red-brown to brown-black		vitreous to resinous	sub-conchoidal fracture	usually twinned at 90° (cruciform) or 60°	STAUROLITE; $FeAl_4Si_2O_{10}(OH)_2$
8.0	colorless, yellow, brown, etc		glassy to resinous	1 directional	transparent to opaque	TOPAZ; $(AlF)_2SiO_4$
9.0	brown to red; sometimes bluish gray		adamantine to opaque	often 1 direction of "false cleavage"	very often hexagonal; SG = 4.0; rubies and sapphires are gem-quality corundum	CORUNDUM; Al_2O_3

Lab 4
Silicate Structures

Name _____

Date _____

Purpose: To understand the chemical structures and atomic arrangements of silicate minerals common in the igneous rocks of Oregon.

Materials:

- Set of 2 sizes of Styrofoam™ balls
- Toothpicks

Background: Silicon atoms (atomic #14) have 3 electron shells containing 2 (inner), 8 (middle), and 4 (outer) electrons. Since each atom would like to have 8 electrons in its outer shell, it is looking to lose 4, gain 4, or share with another atom. In the case of the silicates, it will share electrons with oxygen atoms.

Oxygen (atomic #8) has 2 electron shells with 2 and 6 electrons respectively, so it usually gains 2 electrons when forming ionic bonds, becoming a negatively charged ion (O^{2-}). This is why oxygen often bonds with metals such as Fe and Mg, because they have a tendency to lose electrons and become positive. However, as we now know, oxygen also can share electrons in a covalent bond and this is what it does when it bonds with silicon to form silicate structures.

Instructions: Using page 82 in your textbook as a guide, you are going to attempt to recreate the 5 types of silicate structures with styrofoam™ balls and toothpicks.

Each person in your group should create one independent tetrahedron using the small balls as your silica ions, the larger balls as your oxygen ions, and the toothpicks as your covalent bonds. Remember that silicon has 4 electrons to share so it shares one with each of the 4 oxygens.

1. What is your ratio of silicon to oxygen (Si:O): _____

2. Draw the chemical structure showing the valence (outer shell) electrons.

3. How many electrons are orbiting the silicon: _____
   a. Is it happy? _____

4. How many electrons are orbiting each oxygen: _____

5. Will the oxygen now have a tendency to gain or lose electrons? _____
   a. How many? _____
   b. That makes the overall charge on the silicate ionic compound _____.

6. What is a common mineral formed from independent tetrahedra?_____
   a. What is its chemical formula according to your book? _____
   b. Does it exhibit cleavage? Why or why not?

Now, if you do not already have 6 independent tetrahedra, please create them.
Now use your 6 tetrahedra to create 2 single silicate chains. Each should be 3 atoms long.

7.   How many oxygens do you have left over? _____

8.   What is your new Si:O ratio? _____

9.   Draw the chemical structure showing the valence electrons.

10. How many electrons do each of the oxygens now have in their outer orbital? _____

   a.   Where will the charges be on your ionic compound? _____

   b.   Will it be negative or positive? _____

11. What is a common mineral formed from single chain silicates? _____

   a.   What is its chemical formula according to your book? _____

   b.   Does it exhibit cleavage? Why or why not?

   c.   If so, what kind?

Now, combine your two chain silicates together to form a double chain. Remember to only bond the end atoms, not the middle one, so you end up with what looks a little like a ring. You will have to imagine that, rather than a single ring, this double chain continues on indefinitely.

12. What is your Si:O ratio? _____

13. Does your compound still have a charge? _____

14. What is a common mineral formed from double chain silicates? _____

   a.   What is its chemical formula according to your book? _____

   b.   Does it exhibit cleavage? Why or why not?

   c.   If so, what kind?

Now, join your double chain to that of a neighboring table to create a sheet silicate structure. Answer the questions even if you do not create a sheet silicate.

15.   Will your sheet structure have a charge? _____

16.   What is a common mineral formed from sheet silicates? _____

    a.   What is its chemical formula according to your book? _____

    b.   Does it exhibit cleavage? Why or why not?

    c.   If so, what kind?

Now, carefully attempt to create a network by sharing all the oxygen atoms. You should end up with a Si:O ratio of 1:2 and there will be no charge since each oxygen and each silicon will have 8 electrons in its outer orbital. Answer the questions even if you do not create a network silicate.

17.   What 2 common minerals are considered network silicates?

    a.   Do they exhibit cleavage? Why or why not?

    b.   If so what kind?

Lab 5
Igneous Rocks

Name _____   Date _____

Purpose: In this lab exercise, you will learn to identify some of the most common and important igneous rocks. You will also study the various textures and compositions of igneous rocks as well as explosive potential of magma in order to understand their volcanic origins and tectonic settings. Many of these igneous rocks are common in the Pacific Northwest due to the various forms of volcanism (both recent and ancient) in our area.

Materials:

- Igneous rocks samples
- Glass plate
- Hand lens

Background:

IGNEOUS ROCKS

Igneous rocks, often called "Fire Rocks" (*ignis* = fire), form from the cooling (and subsequent crystallization) of molten rock material, called magma. Magma may cool and solidify at great depths (beneath the surface of the Earth to form intrusive (or plutonic) igneous rocks, or it may cool at the surface as a volcanic eruption or lava flow to form extrusive (or volcanic) igneous rocks. Intrusive igneous rocks tend to have relatively coarse-grained textures because magma beneath the surface cools very slowly, thus allowing ample time for growth of large crystals. On the other hand, extrusive igneous rocks exhibit a much finer-grained texture because they cool rapidly when they are exposed to the cooler temperatures in our atmosphere as lava flows or violent volcanic eruptions. Mineral composition must also be considered when classifying and identifying an igneous rock. Once texture and composition are determined, the rock name can be found pretty easily.

TEXTURE

Because there are a variety of environments in which magmas cool to form igneous rocks, there are a number of different textures that igneous rocks may exhibit. Igneous texture specifically refers to the size, shape, and arrangement of mineral grains within the rock. Because it is a direct consequence of the rate of magma cooling (as well as a few other physical or chemical conditions), a great deal of information about the environment of formation can be learned by carefully studying texture. Below is a discussion of several igneous textures you will observe in this lab exercise.

Phaneritic (Coarse-grained)

Rocks with a phaneritic texture have crystals that are readily visible with the naked eye. This texture is a characteristic of intrusive igneous rocks because it results from large magma bodies that have cooled very slowly (perhaps over hundreds of thousands of years) at great depths (perhaps many kilometers) beneath the Earth's surface. Phaneritic textures consist of a mass of intergrown, equidimensional (roughly equal in size) crystals, which are typically 1-5 mm in size. A hand lens is often used to magnify crystals for identification.

Aphanitic (Fine-grained)

Rocks with an aphanitic texture have crystals that are microscopic and are generally not visible with the naked eye. This texture usually results from fairly rapid cooling at (or just below) the Earth's surface. Mineral composition for aphanitic rocks is typically determined by a polarizing microscope or inferred from color, since properties cannot be determined from such small crystal sizes.

Porphyritic

Rocks with a porphyritic texture have two distinct crystal sizes – larger crystals (phenocrysts) embedded in a finer-grained matrix of substantially smaller crystals (groundmass). Rocks with a porphyritic texture have an interesting cooling history in that they typically represent 2 distinct cooling stages. Because minerals can crystallize under a wide range of temperatures, certain mineral crystals can get quite large before other minerals in the rock have begun to crystallize. If magma that contains some crystals is suddenly shifted to an environment that allows for more rapid cooling (e.g., during a volcanic eruption), the liquid portion of the mixture will crystallize quickly and surround the already formed crystals. A rock with a porphyritic texture is referred to as a **porphyry**. Porphyritic rocks are further classified based on whether the groundmass is phaneritic or aphanitic. If it has a phaneritic (visible crystals) groundmass, it is considered intrusive. If it has an aphanitic (microscopic crystals) groundmass, it is considered extrusive.

Glassy

Rocks with a **glassy** texture actually lack crystals and are considered amorphous ("without shape"). When molten rock is ejected into the atmosphere, it is quenched so quickly that atoms within the magma do not have time to arrange themselves into an orderly crystal structure and are instead "frozen in place." This texture is easily recognizable as solid volcanic glass (e.g., obsidian) or fine shards of intertwined glass (e.g., pumice).

Vesicular

Some fine-grained rocks contain many spherically-shaped voids left behind from trapped gas bubbles. This is often seen near the top of a lava flow because gases in magma rise toward the surface. If gas bubbles can't escape before the lava hardens around them, the resulting volcanic rock is said to have a **vesicular** texture. Occasionally the voids, or **vesicles**, become filled in with another mineral, such as calcite or chert, after formation. If this is the case, the filled-in vesicles are called **amygdaloids**, and the rock name follows the qualifier "amygdaloidal" (e.g., amygdaloidal basalt or amygdaloidal rhyolite).

Pegmatitic

Rocks with a **pegmatitic** texture contain abnormally large crystals (>1cm) and are referred to as **pegmatites**. This texture results from very slow cooling and other physical and chemical characteristics of the magma. Most pegmatitic rocks are similar or identical to granite in composition (see below).

Pyroclastic

Rocks with a pyroclastic texture have fragments of mineral grains and ash that have been welded together by the intense heat of a volcanic eruption. Bits of pre-existing volcanic rock may be visible or microscopic.

COMPOSITION

Surprisingly few minerals make up most igneous rocks. For coarse-grained rocks, mineral composition is relatively easy to recognize, especially with the aid of a hand lens. For fine-grained rocks, however, mineral composition must be indirectly surmised from the color or shade of the rock. Most minerals found within igneous rocks are silicates. Recall that silicates are considered either "light" or "dark" based on the presence or absence of iron and magnesium. Darker-colored igneous rocks tend to contain a great deal of dark, ferromagnesian minerals, whereas lighter-colored igneous rocks do not. Below is a discussion of the igneous compositions which you will observe in this lab exercise.

Felsic

Light-colored igneous rocks are predominantly composed of light silicates, such as feldspar and quartz, and are referred to as **felsic** (fel = *fel*dspar, sic = *si*lica) or **granitic** (from the common crustal rock, *granite* – discussed in intrusive rocks section on next page). Some felsic rocks have 10% or less dark silicates, such as hornblende or biotite mica. Overall, however, most felsic rocks are light in color and have 70% or greater silica content. Continental crust is predominantly composed of granitic igneous rocks.

Mafic

Dark-colored igneous rocks contain mainly dark, ferromagnesian silicates, such as olivine, pyroxene, hornblende, and biotite mica as well as calcium-rich plagioclase (such as labradorite), but no quartz. These are referred to as **mafic** (ma = *ma*gnesium, f = *f*errium [Fe]), or **basaltic** (from the common crustal rock, *basalt* – discussed in the extrusive rocks section on page 33). Mafic rocks tend to be dark in color and greater in density than felsic rocks and contain less than 50% silica.

Oceanic crust is primarily composed of basaltic igneous rock, although it is also common on volcanic islands, and, to a lesser extent, on the continents as lava flows.

Intermediate

Intermediate igneous rocks contain roughly equal amounts of light and dark silicates and are therefore considered somewhat of a "hybrid." The color of intermediate rocks is usually described as medium, or "gray." Rocks of intermediate composition are commonly produced in areas of volcanism as a result of subduction[1]. Volcanoes along the Ring of Fire around the Pacific Ocean tend to be of intermediate composition.

Ultramafic

Ultramafic rocks are almost entirely composed of the ferromagnesian minerals olivine and pyroxene, and therefore are noticeably denser than rocks of other mineral compositions. Ultramafic rocks are relatively rare on the crust, but they are common in Earth's uppermost mantle. Where ultramafic rocks exist on the surface, they have typically either been pushed upward in a tectonic process called obduction[2] or where unmelted parts of the mantle have been brought to the surface by rapidly rising magma during a volcanic eruption. Not all ultramafic rocks are dark in color. Some are a very light green.

IMPORTANT IGNEOUS ROCKS

Intrusive Igneous Rocks:

Granite

Granite is a very well-known igneous rock because it is commonly used in decorative stone. It is intrusive, phaneritic, and must contain both quartz and orthoclase. Accessory minerals such as hornblende, muscovite, biotite, and sodium-rich plagioclase (such as albite) are often present in smaller amounts. Orthoclase typically comprises about 80% of the rock. Granite is light in color overall and tends to be light gray to various shades of pink or red. Some granite is mostly white with some minor amounts of dark silicates. Some granites are further categorized based on color or minor textural differences:

- **Granite Pegmatite** is granite with a pegmatitic texture. It usually has large, irregularly shaped crystals of gray quartz and feldspar. Some contain large "books" of mica (biotite or muscovite).

- **Granite Porphyry** is cooled in 2 distinct stages like other porphyries. Typically it will have a groundmass of medium-sized grains and phenocrysts of larger crystals.

Granodiorite

Granodiorite is coarse-grained and of intermediate to felsic composition. It can look very similar to diorite but should show more light silicates than dark. Likewise, can look similar to "white granite", but shows a greater percentage of dark silicates. Side-by-side comparisons are helpful to distinguish.

Diorite

Diorite is coarse-grained and of intermediate composition. It is usually distinguished by being composed of approximately half light and half dark minerals. It is comprised generally of light colored-plagioclase feldspar, hornblende, and pyroxene, with small amounts of biotite.

---

[1]Subduction refers to the tectonic process of convergence where an oceanic crustal rock is forced beneath either an oceanic plate or a continental plate.

[2]Obduction refers to the tectonic process of convergence where an oceanic crustal plate is thrust onto (above) a continental crustal plate. This is the opposite of subduction and is relatively rare since oceanic crust is denser and thinner than continental crust.

Gabbro

Gabbro is dark, coarse-grained rock with mafic composition. It is predominantly composed of dark, calcium-rich plagioclase and pyroxene, sometimes with small amounts of lighter green olivine. In some gabbros, the mineral labradorite adds a distinct play of colors which make the rock desirable for exteriors of buildings and monuments.

Peridotite

Peridotite is ultramafic and composed entirely of ferromagnesian minerals – usually olivine and lesser amounts of pyroxene. It is typically dark to light green. If it is entirely olivine, it will be a light green color and is referred to as dunite.

Extrusive Igneous Rocks:

Rhyolite

Rhyolite is the extrusive, fine-grained equivalent of granite. Because it contains mostly light silicates, it appears as a light pink, buff, gray, or tan color. Often rhyolite will contain small glass fragments, "flecks" of mica, or voids, which are a result of its rapid cooling. Sometimes voids are filled in with other minerals after its formation. Rhyolite porphyry typically contains light-colored phenocrysts of quartz or feldspar.

Andesite

Andesite is the medium gray, extrusive, fine-grained equivalent of diorite. Its name is derived from the Andes Mountains of South America which are almost entirely made of this rock. Andesite porphyry typically contains dark-colored phenocrysts of hornblende and light-colored phenocrysts of plagioclase feldspar.

Basalt

Basalt is the black, extrusive, fine-grained equivalent of gabbro. It is typically black, very dark gray, or occasionally a dark greenish-black. Basalt makes up the majority of oceanic crust and is the most abundant aphanitic igneous rock on Earth.

Obsidian

Obsidian is compact volcanic glass that lacks crystals. When molten material cools very rapidly, crystals cannot form before the ions lose mobility and the lava solidifies. Obsidian has a distinctly glassy texture, which makes it easy to recognize. Often it will contain dark brown or red streaks. Despite its characteristic dark color, obsidian has a felsic composition and is almost entirely silica. Small amounts of trapped metallic ions give it a dark color. If you can observe obsidian in a fairly thin section (an edge, for example) you can see that it is actually transparent. Because obsidian is dominantly silica, it has a conchoidal fracture and can create very sharp edges (similar to the rock chert).

Pumice

Pumice is a white to dark gray, very light, low density, felsic igneous rock. Like obsidian, it cooled rapidly, but instead of being compact volcanic glass, pumice contains numerous, intertwined shards of glass. This is apparent when it is observed through a hand lens. When hot gases are extruded from an erupting volcano, lava tends to "froth up" and solidify as the pieces are propelled from a volcanic vent. Pumice is often also considered vesicular in texture due to the fact that escaping gas has also played an important role in forming this rock.

Scoria

Scoria is a vesicular rock that forms at the uppermost few feet of a basaltic lava flow where rapid cooling causes the gases within it to migrate upward. Before the gas bubbles escape, the lava hardens around it, leaving behind spherical, tabular, or elongated voids. Fresh scoria is black and glassy, but later often turns reddish brown and dull. Scoria is also sometimes considered glassy as cooling can be too rapid to form crystals.

Tuff

Tuff or welded tuff is an extrusive rock that forms during an explosive volcanic eruption. It is predominantly lithified ash although some larger fragments are sometimes present.

Volcanic Breccia

Like tuff, volcanic breccia is an extrusive rock that forms during an explosive volcanic eruption and contains a significant amount of ash. In volcanic breccia, however, larger angular rock fragments will also be significant.

Lab 6
Understanding Viscosity

Name _____   Date _____

Purpose: In this exercise, you will explore the viscosity of several fluids and observe the influence of temperature on viscosity. You will then apply this understanding to investigate the relationship between the viscosity of magma, its liquid and gas content, and the explosive potential of volcanic eruptions.

Materials:

- Stopwatch
- Metric ruler
- Aluminum foil
- Thermometer
- Calculator
- 12 oz bottle of soda (fresh – not flat)
- Rock ramp
- Protractor
- Corn syrup
- Molasses
- Water

- 3 - 40ml beakers
- Non-vesicular pa'hoe'hoe
- A'a
- Rhyolite
- Andesite
- Scoria
- Pumice
- Obsidian
- Tuff
- Lava Bomb (or picture of)
- Jar of ash

Background: Viscosity is an important property of fluids. It refers to the resistance to flow or deformation of a material (usually a liquid or a gas). Highly viscous liquids are often thought of as "thick" or "sticky," while low viscosity liquids are more "runny." The chemical composition and internal molecular structure of the substance play the largest role in its determining viscosity, but temperature is also very important.

The viscosity of magma at depth influences the nature and style of volcanic eruptions. In magma, viscosity is primarily determined by its temperature and silica content. We will also explore the role that gasses play in magma.

PART 1: VISCOSITY OF A LIQUID

The equation used to calculate viscosity of a liquid is:

$$n = \frac{g \cdot h^2 \cdot p \cdot \sin(A)}{3V}$$

where **V** is the mean velocity of the flow, **g** is the acceleration of gravity (1,000 cm/s²), **A** is the angle of the slope, **h** is the depth of the flowing liquid, **p** is the density of the liquid (in this case, 2 g/cm³), and **n** is the coefficient of viscosity (measured in a poise, which is 1 g/cm s). There is no need to change any units, just plug and chug.

Procedure:

a. Measure the angle of your ramp. _____

b. Place one beaker of corn syrup into the ice bath and place a second one on the hot plate at a low temperature.

c. Create three troughs out of aluminum foil about 10-12 cm in length with a stop at the bottom.

d. Mark a start point and an end point on each trough, 10 cm apart.

e. Measure the temperature of your third beaker of corn syrup.

f. Pour the beaker of room temperature corn syrup onto the start point of your trough and measure how long it takes to reach the 10 cm line. One student should pour while the other times the run. **Try to pour at a steady rate throughout the timing of the run as if the corn syrup is magma being continually being pushed from a volcanic vent.**

g. As soon as your "lava" reaches the 10 cm line, measure the height of the toe of the flow. This will be tricky, so a third student should be ready to go with the ruler.

h. Repeat these steps for each of the other liquids listed in the table below, your hot syrup should be around 30° C, and your cold should be around 10°C. Record the temperature, time, and flow height for each run.

i. Use the data collected in your table to calculate the viscosity of each liquid.

Table 1: Measuring Viscosity

Trail	Temperature	Velocity (cm/s)	Lava Depth (cm)	Angle of Slope	Viscosity (in poises)

Based on your calculations and the plots, answer the following questions:

1. Which liquid had the lowest viscosities?

2. Which temperature had the lowest viscosities?

3. Which liquid had the highest viscosities?

4. Which temperature had the highest viscosities?

5. Does viscosity increase or decrease with temperature?

6. Is the data well correlated? In other words, does it seem like there is a strong relationship between viscosity and temperature, or is your data more scattered?

7. Do you think your data is accurate? If not, why not? Explain where some of the errors might come from in this experiment.

8. How does angle of the slope relate to viscosity?

9. How might the viscosity of a lava flow change with distance from the vent? Explain why in your answer.

PART 2: ROCK SAMPLES

1. Now look at the two rock samples provided at **station 1**. They are both basalt. The ropey, smooth one is called pa'hoe'hoe (pronounced pa-hoy-hoy) and the rough, clumpy one is called a'a (pronounced ah-ah). Based on the two samples' textures, which sample do you think was a higher temperature flow? Which sample has a higher viscosity? Which sample seems to have more vesicles (gas pockets)? What does this suggest about the relationship between temperature and a magma's ability to hold exsolved gases?

2. At **station 2** are samples of rhyolite (pink) and andesite (grey). Describe these two rocks by including their color, mineral composition, and whether or not they show any vesicles (gas pockets). Based on what you know about their composition and what you have learned in class, which lava has the highest viscosity?

3. At **station 3,** grab a bottle of soda. Open it. How explosive was the first opening of the bottle? Now put the cap back on your bottle and shake up the soda. Once you have shaken it, squeeze the sides of the bottle. What has happened to the pressure inside the bottle?

4. Now, slowly uncap the bottle trying to let the gas and not the soda out of the bottle. What has happened to much of the CO_2 (carbonation) of the soda? Is the soda more explosive after shaking?

5. At **station 4** are sample of scoria (reddish black) and pumice (gray). Answer each question for both scoria and pumice (2 answers per question). What type of lava did each of these samples come from? Also, what type of vent do you think each sample came from?

6. **Station 5** is an example of a rock that forms from the collapse of the ash column of an erupting volcano. Explain the texture and perhaps some of the components that you see in the sample. This is what is referred to as a pyroclastic rock. What sort of volcano and what type of lava do you think creates this type of deposit?

7. **Station 6** shows a rock that comes from a rhyolitic lava. What is the name of this rock? Does it have any minerals present? Explain why or why not? This sample is very similar to pumice, but what is the difference? Is the sample mafic or felsic?

8. The sample at **station 7** is called a lava bomb. How does something like this volcanic sample form? What type of rock is it?

9. At **station 8** are containers of ash from the Mt. St. Helens eruption. In large volcanic eruptions from andesite and rhyolite volcanoes, particulates like ash and chemicals like sulfuric acid can get into the stratosphere where there is no wind or weather, like rain. Shake up these containers and look through to see the contents. What are the possible outcomes of having lots of stagnant ash in the stratosphere (The stratosphere is the layer of the atmosphere that contains the ozone layer)?

10. Based on what you have seen, how are temperature, gas content, and viscosity connected when determining the explosivity of an eruption? Based on these 3 factors, what type of lava would have the lowest viscosity? Which lava would have the highest viscosity? What type(s) of lava would be the most dangerous to live by?

11. Rate the next three volcanoes in terms of danger during eruption; Yellowstone Volcano, Mt. St. Helens, and Mauna Loa, Hawaii's volcano. Refer to what you have learned in your answer. What lava is associated with each volcano? What sort of threats are associated with each lava?

Lab 7
Volcanic Hazards

Name _____   Date _____

Purpose: In this lab exercise, you will explore the volcanic hazards that could affect areas around
Mt. Hood in the event of an eruption. You will also be assigned the role of a USGS representative for
a given volcanic scenario.

Materials:

- Mt. Hood South Quadrangle

PART 1:

Use the Mt. Hood South topographic map to answer the following questions:

1. Name 6 of the main rivers or creeks draining off of the volcano and describe the general direction in which they are flowing.

2. Count and name the glaciers on Mt. Hood. What percentage of the mountain is covered in glaciers (make a good estimate here)?

3. Are there any towns or villages on the map? How far away from the peak of the volcano are they?

PART 2

In this section, you will work in a group acting as the US Geologic Survey (USGS). Once you are given information regarding the activity of a local volcano, you will have to decide what should be done with regard to whether increased studies with GPS and field work should be ordered, how much and what information should be released to the press, and what should be done to evacuate local people or help them to prepare for an impending eruption.

1. At what point, if at all, did you begin further monitoring the volcano? What sort of tools did you use for monitoring?

2. At what point did you alert the press, if at all, about the issues concerning the volcano? What sort of information did you share?

3. At what point, if at all, did you evacuate local areas? What areas did you decide to evacuate and were you correct about your evacuation radius?

4. Based on the ash distribution and lahars formed during the eruption, were the steps you and your classmates took during the evolution of volcanic activity enough to protect the Oregonians near Mt. Hood? What, if anything, would you have done differently?

Lab 8
Hollywood Geology

Name _____   Date _____

Purpose: To use the knowledge gained during this class to find both the correct and incorrect geologic information as portrayed by Hollywood movies.

Instructions: Your instructor has selected a Hollywood movie that depicts geologic events. Your job while we watch this movie is to take detailed notes about the events that you see. You will then write a 2-page paper in double spaced, 12-point font, about how accurate the movie was using actual evidence (scenes) from the movie. What about the movie was totally bogus based on what we have learned in class? What in the movie is accurate or plausible based on what we have learned in class? If you have any questions, ask your instructor.

Notes:

3. Describe the unit Trh.

4. Go to the pin in your .kmz file called Leslie Gulch. This is made of Trh. How do the rocks of the Steens Mountain and Leslie Gulch compare? These two features are an example of bimodal volcanism from the same source. What do you think the source of these rocks are?

5. Go to the pin in your .kmz file for the Wallowa Mountains. Are these volcanoes? Explain why or why not.

6. Go to the pin in your .kmz file called Mary's Peak. Notice that the top of the peak is flat. The rock that creates the peak is volcanic. What type of rock do you think creates the peak?

7. Go to the geologic map and find Mary's Peak. Describe the geologic unit that you find there. Were you correct in your answer for question 6?

8. Go to the pin in your .kmz file called Mt. Ashland. Now find it on the geologic map. What is the unit that composes Mt. Ashland? What other pin you've seen has shown similar geology?

9.   Go to the pin in your .kmz file called Mt. Tabor. What type of feature is this?

10.   Now go to the geologic map and find Mt. Tabor. What is the rock that composes this feature? Was your answer correct in question 9?

11.   Go to the pin in your .kmz file called Three Sisters and then go to the pin in your .kmz file called Mt. Bachelor. Compare these two different volcanoes.

12.   Go to the pin in your .kmz file called Fort Rock. Describe this feature.

13.   Now go to the geologic map and find Fort Rock. Describe the geologic unit. What does subaqueous mean?

Lab 10
Profiling Mt. St. Helens

Name _____   Date _____

Purpose: In this lab exercise, you will analyze the change in shape, elevation, and gradient of Mount Saint Helens after its 1980 eruption. You will also view a PBS documentary detailing the 1980 eruption, subsequent activity, and its "return to life" from a biological perspective.

Materials:

- ◆ Ruler or straight edge
- ◆ Pencil with eraser
- ◆ Calculator

Background: Mount Saint Helens (MSH) is one of 13 composite cone volcanoes that together make up the most prominent peaks of the Cascade Range, which extends from northern California into Canada. MSH, in southern Washington state, is currently the most active member (figure 1).

The Cascade volcanoes are a result of the convergence of the Juan de Fuca plate (oceanic crust) below the North American plate (continental crust) where subduction of the Juan de Fuca plate heats the descending lithosphere causing partial melting and magma production. This results in periodic lava flows and pyroclastic eruptions. The lava is typically andesitic or dacitic (between andesite and rhyolite) in composition.

On May 18th, 1980, MSH suddenly erupted, killing dozens of people. It represented the deadliest and most explosive volcanic event in recorded US history. The explosion was unique in that it was directed laterally (sideways) and to the north. The volume of ejecta is calculated to be around 0.3 cubic miles (1.25 cubic kilometers) which ranks it at a VEI 5 (barely). See figure 2.

The following exercise is designed to let you view cross-sections of the volcano before and after 1980 by means of topographic profiling and calculating elevation and volume changes.

Figure 1. Major Cascade Range Volcanoes (MSH shown in red).

Figure 2. Digital map of MSH before (above) and after (below) 1980 eruption.

DRAWING A TOPOGRAPHIC PROFILE

A **topographic profile** is essentially a cross-sectional view of a map area along a specific line. Topographic profiles can be helpful in that they can visually show the relief of an area of interest. Follow the steps below.

1. Locate the transect line on the map. Place a blank piece of paper along the line and mark the starting and ending points of the line. Indicate where the transect begins and ends. Usually "A" marks the start and "B" marks the finish.

2. Starting at one end, move along the edge of the paper, making a mark on the paper every time a contour line touches the edge of the paper (figure 3). Make sure you label each mark with the right elevation so that you can transfer that point to the correct elevation on your profile.

3. For every mark you made on the paper, place a vertical point on the graph directly above that point corresponding to the correct elevation (figure 4). Use edge of paper as a square.

4. Connect the dots on the graph paper to complete topographic profile.

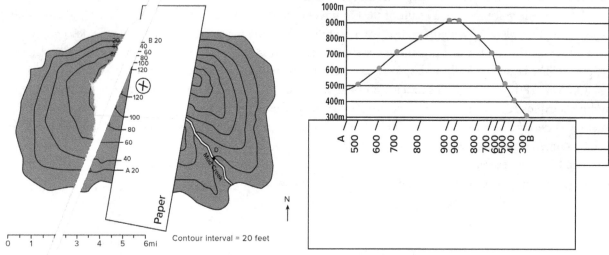

Figure 3. Making a transect line.

Figure 4. Creating a profile on a graph.

5. Repeat this process for the following profiles:

   a. Using the pre-eruption map of MSH (figure 5), complete an A – B profile (figure 6).

   b. Using the post-eruption map of MSH (figure 7), complete an A – B profile (figure 8).

   c. Using the pre-eruption and the post-eruption maps of MSH, complete an X-Y profile for each (figure 9).

Lab 11
Earthquakes

Name _____   Date _____

Purpose: In this four-part exercise, you will determine the Richter magnitude and location of an actual earthquake epicenter, calculate the recurrence interval of earthquakes in the Pacific Northwest, and investigate the relationship of plate tectonics to global seismic activity.

Materials:

- Ruler/straight edge
- Colored pencils
- Compass
- Calculator

Background: Over 90% of the world's earthquakes and volcanoes occur at or near plate boundaries. Those earthquakes that occur within plates are referred to as intraplate earthquakes. These are typically caused by a slip along an ancient fault. The study of earthquakes and earthquake waves is a branch of geophysics called seismology. An earthquake is essentially the shaking or trembling of the Earth's crust. They are generated when rock beneath the Earth's surface suddenly ruptures or by volcanic activity. Shock (seismic) waves are sent off from the region of rupture in all directions. The earthquake hypocenter (or focus) is the point below the Earth's surface where these seismic waves originated during rock rupture. Earthquake epicenters are classified according to their depth from surface to hypocenter.

> **Shallow:** up to 70km (45 miles)
>
> **Intermediate:** 70 – 300km (45 – 190 miles)
>
> **Deep:** Greater than 300km (186 miles)

PART 1: LOCATING THE EPICENTER AND FINDING RICHTER MAGNITUDE

Procedure: For each seismic station in Table 1, label the P-waves and S-waves on each of the seismograms in figure 1. Then, determine the arrival time of the P-waves and S-waves. (Note: The x-axis of each seismogram on the next page has been adjusted so that P-wave arrival time is at 0 seconds.) With this information, you can determine the distance to the epicenter by using the graph (figure 2). Draw a horizontal line extending from the y-axis at the value corresponding to the time interval, stop where it intersects the line, and draw a vertical line from that point to the x-axis to determine the value for distance.

Once you have determined distances for all three seismic stations, use a compass to find the epicenter of the quake on the map in figure 3. Set your compass to the appropriate distance according to the scale on the map. Place the point at the station site and inscribe one arc with the pencil. Follow this procedure for each station. Once this is complete, trace each of the circles with colored pencil: green for Eureka, red for Elko, and blue for Las Vegas. The point where all 3 arcs intersect is the epicenter. Indicate the epicenter with a big yellow dot. Note: Do not be too concerned if the intersection is not exactly a single point, but it should be close.

Finally, determine the maximum S-wave amplitude for each seismogram with Table 2 and plot the amplitude/distance to epicenter values for each station on the nomogram provided in figure 4.

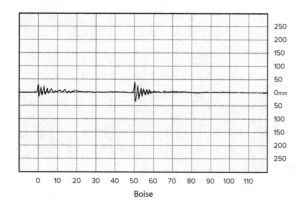

Boise

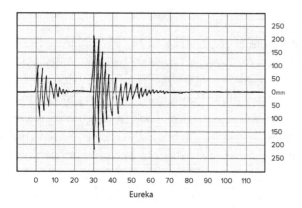

Eureka

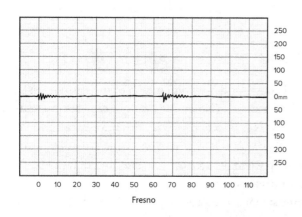

Fresno

Figure 1. Three seismograms representing the seismic record of same earthquake at three different locations.

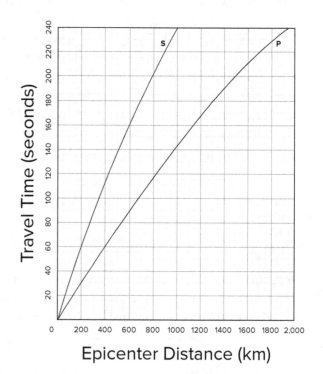

Epicenter Distance (km)

Figure 2. The S-P interval with the time difference between the arrival times of the first P and first S waves (Y-axis) and corresponding epicenter distances (X-axis).

Table 1

	S-wave arrival time (sec)	P-wave arrival time (sec)	S-P Interval (sec)	Distance from Epicenter (km)
Eureka, CA				
Elko, NV				
Las Vegas, NV				

Figure 3. Map of western United States with locations of seismic stations.

1. What is the Richter magnitude of this earthquake? _____

2. Which of the following best fits the location of the earthquake epicenter?

   a. between Salt Lake City and Provo, UT

   b. in the Pacific Ocean, southwest of San Francisco, CA

   c. between Fresno and San Francisco, CA

   d. directly east of Klamath Falls, OR

   e. along the Mexico – California border

   f. way off the map…I think I did something wrong

3. Which of the following cities should receive the earthquake waves <u>first</u>?

   a. San Francisco, CA

   b. Bakersfield, CA

   c. Ogden, UT

   d. Tonopah, NV

   e. Boise, ID

4. Which of the following BEST fits the distance, in kilometers, from the city of Los Angeles to the epicenter?

   a. 120

   b. 300

   c. 500

   d. 700

   e. 1300

Table 2

	Maximum S-wave Amplitude (mm)
Eureka, CA	
Elko, NV	
Las Vegas, NV	

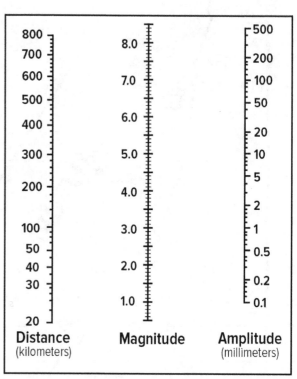

Figure 4. Nomogram for determining Richter magnitude (ML). Use designated colors to draw lines for each seismic station (green for Eureka, red for Elko, and blue for Las Vegas).

PART 2: CASCADIA RECURRENCE INTERVAL

Perhaps you have heard on the radio, read in the newspaper, or watched on TV that the Cascadia subduction zone is gearing up for a very large earthquake. This would pose a serious threat to coastal communities and large urban areas like Portland, Seattle, and Vancouver, B.C. But is it possible to determine how likely it is that a large earthquake will strike the Pacific NW? You and your lab partners will use **conditional probabilities** (which is the likelihood that a specific event will occur, given that another event has already occurred), to estimate the probability of an earthquake of a given magnitude occurring in an area within a certain time period. These estimates are based on real data collected from the US Geological Survey (USGS) and historical records from the Pacific Northwest Seismic Network (PNSN). In this activity, you will analyze the seismic history of earthquakes in the Pacific Northwest to estimate the hazards we might be facing in the future.

Procedure: Using the data below, complete the following table using the instructions below and answer all questions regarding the data you have been given and calculated.

Table 3

Two-year period	Number of M 2.0-2.9 earthquakes
1992-1993	113
1994-1995	62
1996-1997	59
1998-1999	57
2000-2001	49
Total in ten-year period:	
Three-year period	**Number of M 3.0-3.9 earthquakes**
1988-1990	52
1991-1993	151
1994-1996	84
1997-2000	108
2001-2003	68
Total in 15-year period:	
Five-year period	**Number of M 4.0-4.9 earthquakes**
1979-1983	151
1984-1988	39
1989-1993	45
1994-1998	31
1999-2003	29
Total in 25-year period:	

For the questions, you should know the following information about magnitude 5 through 7 earthquakes:

Table 4

Magnitude and # of years	Number of Earthquakes
5.0-5.9, 1973 - 2003 (30)	25
6.0-6.9, 1872 - 2003 (131)	9
7.0-7.9, 1872 - 2003 (131)	3

Questions

1. What are some problems you might see with the data above? How significant are these errors? (Hint: think about how averages work.)

2. Look at your M 4.0-4.9 data table. Between March 24 and May 19, 1980, there were over 100 earthquakes between M 4.0-4.9. Why is this?

3. Should you include 1980 data in your calculations or not? Why or why not? Fully discuss your answer.

4. Fill out the following table using the instructions below:

   a. To calculate the average number of earthquakes in 1 year **divide the total number of earthquakes** for each magnitude (calculated in tables 3 and 4) **by the number of years in your data set.** (Column 2 / Column 3)

   b. To calculate the expected number in ten years by **multiplying the average number of earthquakes expected in one year by 10.** (Column 4 x 10)

   c. To calculate the time between earthquakes in days **divide 365 days per year by the average number of earthquakes in one year.** (365 / Column 4)

   d. To calculate the time between earthquakes in years **divide time between earthquakes in days by 365 days.** (Column 6 / 365)

   e. To determine whether your calculations are correct, think about the last column: time between earthquakes in years. Are the years between earthquakes increasing? They should be. How many magnitude 7 earthquakes have you experienced here in Oregon?

Table 5

Magnitude range	Number of earthquakes (calculated above)	Years in data set	Average number in one year	Expected number in ten years	Time between earthquakes (days)	Time between earthquakes (years)
2.0-2.9		10				
3.0-3.9		15				
4.0-4.9		25				
5.0-5.9		30				
6.0-6.9		131				
7.0-7.9		131				

5. Plot the magnitude versus time between earthquakes in years (column 7) on the logarithmic graph paper provided. Plot all magnitude 2.0-2.9 earthquakes as a point on the vertical line labeled 2, all magnitude 3.0-3.9 earthquakes as a point on the vertical line labeled 3, etc.

6. Project how often a M 8-9 earthquake might happen by extending a best-fit line through magnitude 8 and 9 on your graph. What is the average recurrence range of earthquakes of this magnitude?

7. Geologic evidence suggests there was a M 8-9 earthquake about 300 years ago and about 1100 years ago. Does this evidence fit with your calculations?

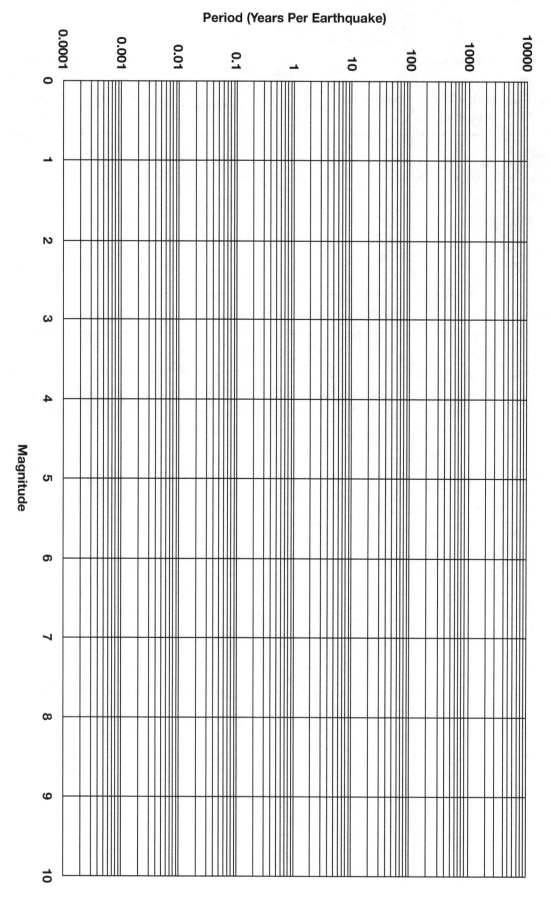

Period (Years Per Earthquake)

0.0001   0.001   0.01   0.1   1   10   100   1000   10000

Magnitude

0   1   2   3   4   5   6   7   8   9   10

Magnitude vs. Period

8. Given the information above, discuss the following question: Should public officials, emergency planners, engineers, and individuals focus on earthquake preparedness efforts for M 8-9 earthquakes or M 6-7 earthquakes? Explain your reasoning in detail. Please write at least one paragraph.

PART 3: EARTHQUAKES AT PLATE BOUNDARIES

Look at the earthquake depth and distribution map (figure 5) and compare it to your volcanic activity map. Answer the following questions.

1. Most earthquakes worldwide are of what depth? (Circle one)

   a. Shallow

   b. Intermediate

   c. Deep

   d. Very deep

2. In general, what depth of earthquakes do you see associated with each of the following?

   a. Mid-Ocean Ridges _____

   b. Intra-plate regions _____

3. The deepest earthquakes are found primarily at what type of plate boundaries? (Circle one)

   a. Divergent

   b. Convergent – subduction

   c. Convergent – collisional

   d. Transform

4. Explain your answer above. Why do you think the deepest earthquakes are found here?

5.   Look closely the Pacific Northwest and the depth of earthquakes here. Is this typical of a subduction zone? Explain this seeming contradiction.

SCIENTIFIC SPECIALTY: SEISMOLOGY

Earthquake Locations 1990 - 1996 (Magnitudes 4 and greater)

Color indicates depth: Red 0-33 km, Orange 33-70 km, Green 70-300 km, Blue 300-700 km

This map is part of "Discovering Plate Boundaries," a classroom exercise developed by Dale S. Sawyer at Rice University (dale@rice.edu). Additional information about this exercise can be found at http://terra.rice.edu/plateboundary .

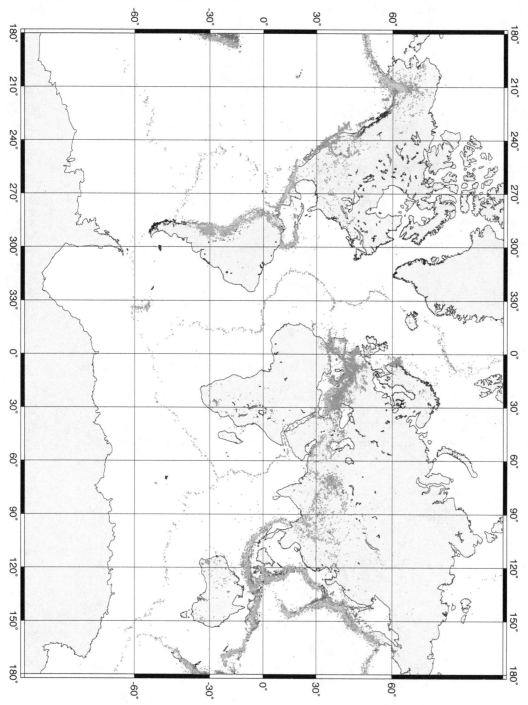

Figure 5. Earthquake depth and distribution map. Courtesy of Dale Sawyer and Rice University. (http://plateboundary.rice.edu/downloads.html).

SCIENTIFIC SPECIALTY: VOLCANOLOGY

Red dots indicate currently or historically active volcanic features

This list obtained from the Smithsonian Institution

This map is part of "Discovering Plate Boundaries," a classroom
exercise developed by Dale S. Sawyer at Rice University (dale@rice.edu)
Additional information about this exercise can be found at
http://terra.rice.edu/plateboundary .

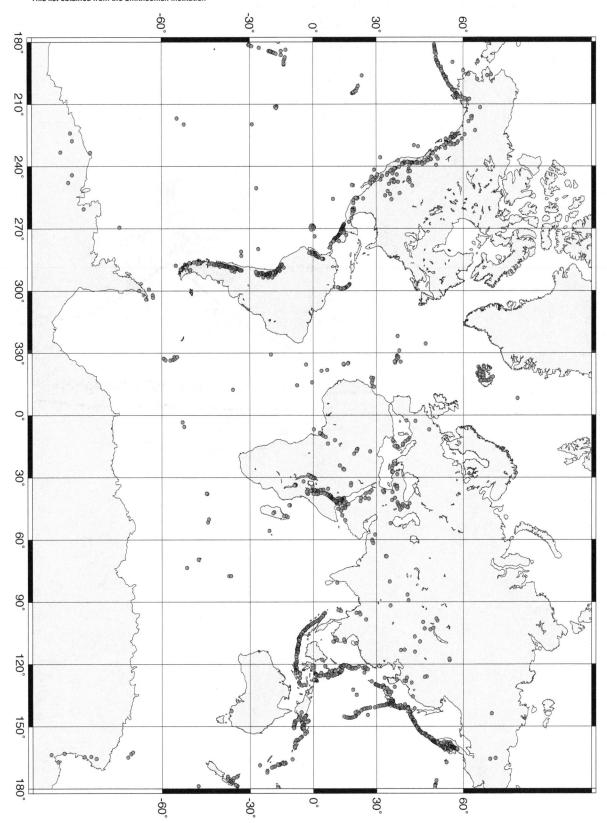

Figure 6: Volcanic activity map. Courtesy of Dale Sawyer and Rice University.
(http://plateboundary.rice.edu/downloads.html).

PART 4: INVESTIGATING THE WADATI-BENIOFF ZONE

Where tectonic plates converge and one or both plates is oceanic lithosphere, a subduction zone will form where an oceanic plate sinks into the mantle. Three major features are associated with subduction zones: (1) a deep ocean trench, (2) a volcanic island arc composed of andesite volcanoes (island arcs) and (3) a plane of earthquake hypocenters shallow near the trench and descending beneath and beyond the volcanic arc. This plane, known as a Wadati-Benioff Zone, typically slopes downward at an angle of about 45° (figure 7).

The Tonga Trench is the second deepest trench in the world, after the Mariana. It is located in the South Pacific Ocean at the convergence of the Indo-Australian and Pacific tectonic plates. In this section, you will locate the approximate plane of subduction associated with this plate boundary by plotting several known earthquake hypocenters.

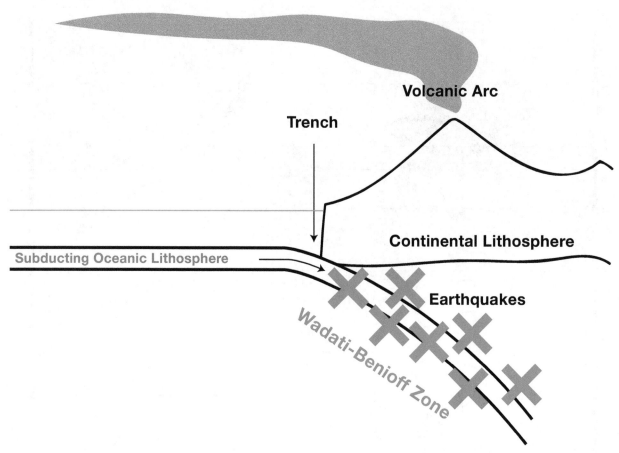

Figure 7. Wadati-Benioff Zone.

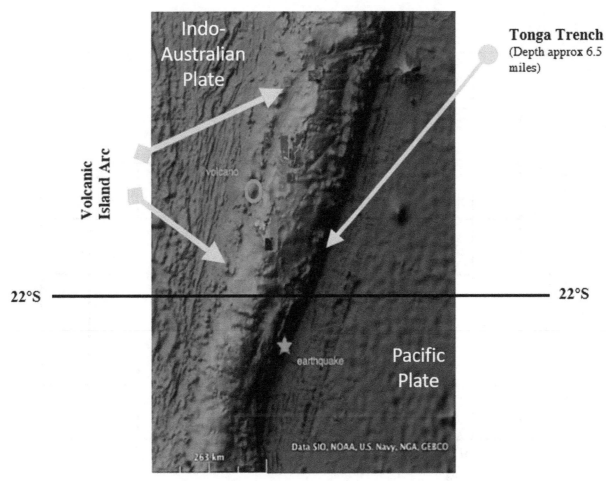

Figure 8. Relief map of Tonga trench.

Procedure: Using the information in Table 6 and the cross section across latitude 22°S (figure 9), plot the location of each earthquake hypocenter using a small red "x."

Table 6: Earthquakes associated with the Tonga Trench

Depth of Hypocenter (km)	66	83	116	150	216	250	300	350	383	448	500	516	550	600	624	667
Distance from Trench Axis (km)	132	184	232	260	264	264	360	400	408	448	450	456	500	508	524	548

72   Geology 142

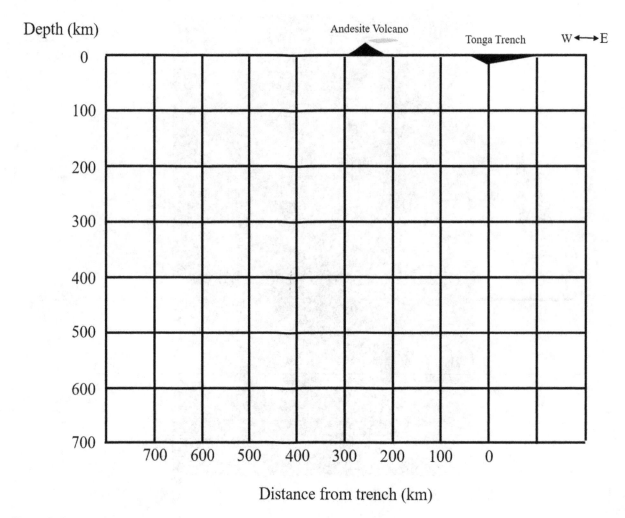

Figure 9. Cross section of latitude 22° S across the Tonga Trench. (graph is not to scale).

Questions

1. Note the Andesite volcano, part of an island arc of volcanoes, that parallels the axis of the trench. Approximately what is the distance, in kilometers, from the axis of the trench to the volcano?

2. The Andesite volcano formed when lithosphere material partially melted, descended into the mantle, and then rose to the surface. From approximately what depth below the surface did the magma begin its ascent? Is this a shallow, intermediate, or deep hypocenter region?

3. Label the 2 plates on your cross section (figure 9).

4. On which plate do the andesite volcanoes lie?

5. Which plate descends into the mantle?

Lab 12
Accreted Terranes of the Pacific Northwest

Name _____   Date _____

Purpose: In this lab you will be creating a map based on the major accreted terranes of the western coast of North America, mainly in Oregon.

Materials:

- ◆ Glue Stick
- ◆ Scissors
- ◆ Colored Pencils

Background: The thing about accreted terranes is that, geologically, they are a mess. They have everything you could possibly imagine in them: sedimentary rocks, igneous rocks, metamorphic rocks, continental rocks, oceanic rocks, fossils, etc. In fact, they are so messy that the geology is usually called a mélange (the French word for mixture). Today you are going to create your own map of these Pacific Northwest accreted terranes. You will also determine the type of accreted terrain based on rock assemblages.

PART 1:

1. You will need to assemble your map by cutting out your accreted terranes and glueing them to the map based on the borders shown and the cities on the terranes and the main map.

2. **After** your terranes are glued onto the main map you will color them based on their age and names (anything with a * is a continental igneous province associated with the accretion of a super terrane):

   a. IMT—Intermontane Superterrane: Middle to late Jurassic Accretion **BLUE**

   b. OB*—Omineca Belt: Middle to late Jurassic Accretion **RED**

   c. KT—Klamath Terrane: Late Jurassic to Early Cretaceous **GREEN**

   d. CPC*—Coast Plutonic Complex: Middle Cretaceous **YELLOW**

   e. EOT—Eastern Oregon Terranes: Middle Cretaceous **PURPLE**

   f. INT—Insular Superterrane: Middle to late Cretaceous **LIGHT BLUE**

   g. OT—Olympic Terrane: Tertiary **ORANGE**

   h. CT—Coastal Terrane: Tertiary to present **BROWN**

3. Using the attached geologic timeline, what is the approximate time of accretion for each of the terranes and volcanic arcs? Where appropriate, use a range of years (ex: 90-40 million):

   i. Intermontane Superterrane: _____

   j. OB*—Omineca Belt: _____

   k. KT—Klamath Terrane: _____

   l. CPC*—Coast Plutonic Complex: _____

   m. EOT—Eastern Oregon Terranes: _____

   n. INT—Insular Superterrane: _____

   o. OT—Olympic Terrane: _____

   p. CT—Coastal Terrane: _____

PART 2:

Use the following rock assemblages to tell me what type of terrane each of the 7 terranes you mapped are composed of. A description is given for each type of terrane first, named A, B, C, or D. After each rock assemblage for the following 7 terranes, there is a blank space to list the type of terrane(s) related to it. Note: sometimes these terranes occur together, so there can be more than one answer, with the exception of the continental volcanic arc.

    a.   Ophiolite: a section of oceanic crust that can contain (from top to bottom): ultramafic rocks, gabbro, sheeted dikes, pillow basalts, siliceous ooze

    b.   Volcanic Island Arc: erupted volcanic rocks intruded by numerous diorite, granodiorite, and granite bodies intruded into the island arcs

    c.   Oceanic Plateaus: some rocks are limestone formed from coral reefs in a marine environment; also contains submarine fans similar to those currently being formed on the continental slope off the west coast

    d.   Continental Volcanic Arc: extensive folding, faulting, and metamorphism due to convergence of an oceanic plate, widespread intrusion by granites

1. **Intermontane Superterrane** contains erupted volcanic rocks intruded by granitic rocks and packets of limestone and coral fossils. _____

2. **Omineca Belt*** consists of a wide spread area of mainly of granite and gneiss (a metamorphic rock consisting of dark and light mineral bands). _____

3. **Klamath Terrane** contains basalt and andesite intruded by granites, limestone pockets, coral fossils, sheeted dikes, and pillow lavas. _____

4. In the **Coast Plutonic Complex*,** massive amounts of molten granite injected into this area metamorphosed local country rock into a glittering medium-grade metamorphic rock called schist. _____

5. **Eastern Oregon Terranes** consists of volcanic rocks like andesite, intruded by numerous diorite, granodiorite, and granite bodies. Terrane also contains sequences of volcanic sediments with limestone and fossils of coral reefs. These aforementioned sequences are sandwiched between pillow basalts, gabbros and sheeted dikes. _____

6. **Insular Superterrane** contains erupted volcanic rocks and is intruded by granitic rocks and packets of limestone and coral fossils. _____

7. **Olympic Terrane** contains a lot of basalt, pillow basalts, and submarine fans similar to those currently being formed on the continental slope off the west coast. _____

8. **Coastal Terrane** contains gabbro and a large amount of basalt. Terrane also contains submarine fans similar to those currently being formed on the continental slope off the west coast. _____

PART 3:

1. In some of the accreted terranes of the Pacific Northwest there are fossils of fusulinids, a type of primitive, single-celled animal that floated in ocean water. They flourished in tropical oceans during the late Paleozoic era, during the Mississippian through Permian periods. The fusulinid fossils in terranes of the Northwest are called Tethyan fusulinids because they are types of fusulinids that existed in the large sea known as the Tethys Sea on the east side of the supercontinent Pangaea. Based on this information and the map (figure 1), how did the terranes of the Pacific Northwest get to their current location.

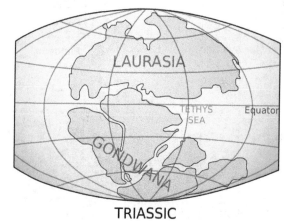

TRIASSIC
200 million years ago

Figure 1. Landform during Triassic Era.

2. There is a line that starts in northeastern Washington and then travels approximately along the border of Washington and Idaho and the border of Oregon and Idaho. This line separates two areas that have different values of Strontium 87. Strontium 87 is an isotope that is stable, meaning it is no longer radioactive. Its parent material is Rubidium 87. Rocks east of this line have a much higher concentration of Strontium 87, while rocks west of this line have a very low concentration of Strontium 87. Explain why this occurs.

3. If you are walking on an ophiolite complex, it is a sure sign that you are walking on accreted terrane.

    a. How do you know you are walking on an ophiolite?

    b. It is rare to see an entire ophiolite sequence in an accreted terrane. Give two reasons for why this is and explain your answer.

ERA	PERIOD		EPOCH		Ma
Cenozoic	Quaternary		Holocene		0.011
			Pleistocene	Late	0.8
				Early	2.4
	Tertiary	Neogene	Pliocene	Late	3.6
				Early	5.3
			Miocene	Late	11.2
				Middle	16.4
				Early	23.0
		Paleogene	Oligocene	Late	28.5
				Early	34.0
			Eocene	Late	41.3
				Middle	49.0
				Early	55.8
			Paleocene	Late	61.0
				Early	65.5
Mesozoic	Cretaceous		Late		99.6
			Early		145
	Jurassic		Late		161
			Middle		176
			Early		200
	Triassic		Late		228
			Middle		245
			Early		251
Paleozoic	Permian		Late		260
			Middle		271
			Early		299
	Pennsylvanian		Late		306
			Middle		311
			Early		318
	Mississippian		Late		326
			Middle		345
			Early		359
	Devonian		Late		385
			Middle		397
			Early		416
	Silurian		Late		419
			Early		423
	Ordovician		Late		428
			Middle		444
			Early		488
	Cambrian		Late		501
			Middle		513
			Early		542

					Ma
Precambrian	Proterozoic	Late	Neoproterozoic (Z)		1000
		Middle	Mesoproterozoic (Y)		1600
		Early	Paleoproterozoic (X)		2500
	Archean	Late			3200
		Early			4000
	Haydean				

Phanerozoic

Figure 2. Geologic timeline.

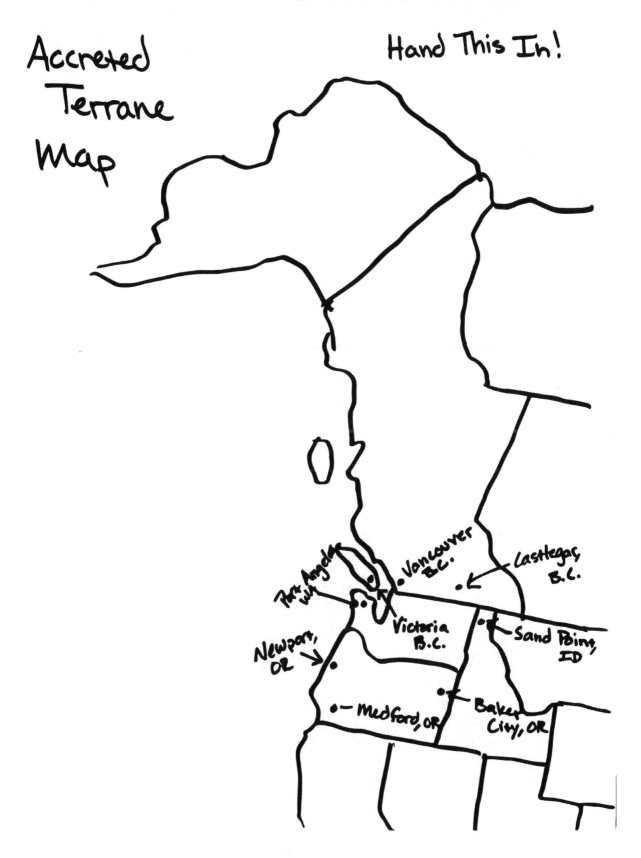

Figure 3. Accreted Terrane map.

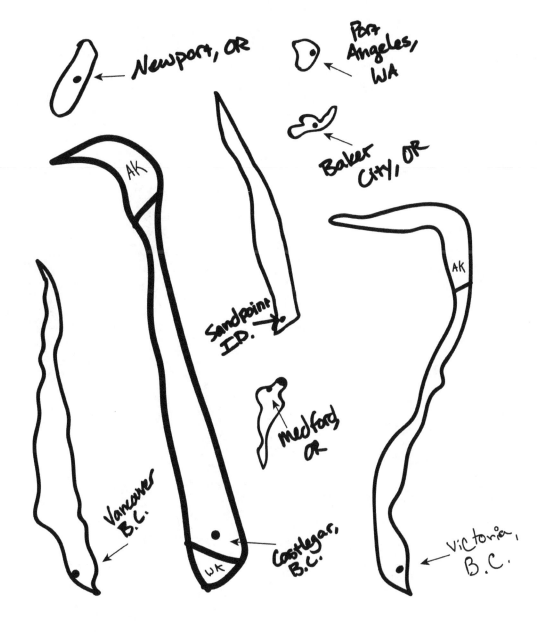

Accreted Terranes: Cut & Color These.

Newport, OR

Port Angeles, WA

Baker City, OR

AK

Sandpoint ID.

Medford OR

Vancouver B.C.

Castlegar, B.C.

Victoria, B.C.

AK

Figure 4. Accreted Terranes.

Lab 13
Geologic Structures

Name _____   Date _____

Purpose: In this exercise, you will explore geologic maps, which are the primary means of communicating geologic information. You will investigate the primary types of crustal deformation (folds and faults), the types of stresses that create them, and how they are represented on geologic maps and cross sections.

Background:

Important Terms:

Geologic map: shows the distribution of rocks at the Earth's surface.

Geologic cross section: a drawing of a vertical slice through Earth, like a cutaway view. It shows the arrangement of rock formations and contacts. A cross section also shows the topography of the land's surface.

Block diagram: a combination of a geologic map and cross section. It looks like a solid block, with a geologic map on top and a geologic cross section on each of its visible sides. A block diagram is like a small 3D model of a portion of the crust.

CROSS-SECTIONS AND MAP VIEWS

Measurement of Orientation

Strike:

- ◆ Compass direction of the outcrop
- ◆ The line formed by the intersection of a horizontal plane with the structure

Dip:

- ◆ The angle between the horizontal plane and the planar surface being measured is always 90 degrees (or perpendicular to strike).

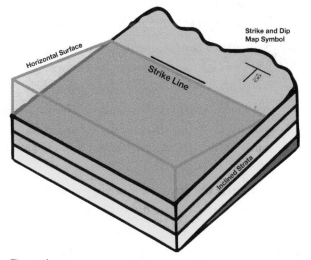

Figure 1.

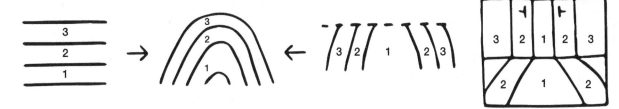

1. Deposition of strata

2. Apply compression and form structure

3. Erosion

4. Resultant outcrop pattern

Plotting Strikes and Dips on Maps

One of the benefits of measuring the strike and dip is to be able to place information on a geologic map about the angle or tilting planar geologic features like bedding and faults. A geologist represents the strike and dip of bedding planes using the symbols shown below.

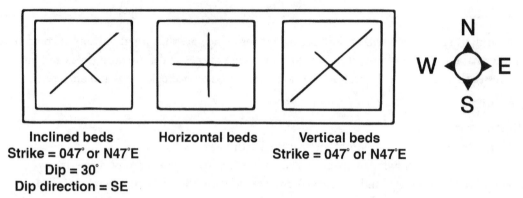

Inclined beds
Strike = 047° or N47°E
Dip = 30°
Dip direction = SE

Horizontal beds

Vertical beds
Strike = 047° or N47°E

How to Make a Cross Section

1. **Make dots on the cross section line:** On a piece of scratch paper, place a dot on the map cross-section line every place that a contact or a fault crosses it.

2. **Transfer dots to the top of the cross section:** The spacing of the dots on the cross-section line need to be carefully transferred to the land surface (top) of the cross section.

3. **Make notes above the cross section:** Using the data on the geologic map, lightly write in the rock layer ages (above the cross section) between the appropriate dots. These numbers, which should be erased after the cross section is complete, help you visualize the geometry below the surface later on.

4. **Burrow the dots into the ground:** The dots at the top of your cross section represent spots where a contact of a fault slants into the ground at some sort of angle. Based on the type of fold present or the fault's geometry, "burrow" the dot into the ground by drawing a short line into the subsurface at the appropriate angle.

5. **Complete the cross section:** The last step is to continue the "burrows" down through the cross section. Remember to draw the rock layers with a constant thickness

Folds:

One of the ways a rock responds to forces is by bending. If you hold up this lab manual and squeeze it, pushing on both ends, it will buckle and bend, forming folds. The same thing will happen to any layered material — when it gets squeezed end-on, it will form folds. Note that if you try to squeeze this manual the other way (pushing straight down on the pages), you will *not* cause it to bend. Therefore, in order to get folds, you need to have some sort of end-on squeezing force applied to a layered material. Since the most common type of layering in rocks is sedimentary bedding, which starts out horizontal, the type of force that most often causes folding in the Earth's crust is compression.

Figure 2. Anticline.

Figure 3. Syncline.

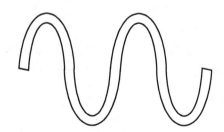

Figure 4. Alternating anticlines & synclines.

There are two basic types of folds. If the beds are bent up into an arch, we call it an **anticline** (think of the letter "A"). If they are bent downward into a trough, we call it a **syncline** (think of a sink). These two types of folds are most commonly seen together, alternating anticline, syncline, anticline, etc.

In nature we most often see a two-dimensional representation of the three-dimensional structure of a fold, as we can only observe what is on the surface of the Earth. Drill holes and sophisticated geophysical techniques can give us information about subsurface geology, but they are expensive and only tell part of the story. Thus, all we readily have access to is the two-dimensional surface that is the result of erosion cutting through the structure. Our job as geologists is to figure out what the whole structure looks like. It's like trying to figure out what a whole head of cabbage looks like when all you can see is the cut surface of it.

Deformation	Shape	Origin	Strike and Dip Pattern
Monocline	A single-limbed bend in rock strata	Compressional forces; vertical motions deeper within the crust	
Anticline	An upward bend in rock strata	Compressional forces	
Syncline	A downward bend of rock	Compressional forces	
Dome	A circular upward bend of rock strata	Common within the continental interiors, Compressional forces	

Figure 5. Some common types of folds with characteristic strike and dip patterns.

Deformation	Shape	Origin	Strike and Dip Pattern
Basin	A circular bowl-shaped bend	Common within the continental interiors	
Plunging anticline	A non-horizontal fold axis	Compressional forces; can be formed by two tectonic events	
Plunging syncline	A non-horizontal fold axis	Compressional forces; can be formed by two tectonic events	

Figure 5. Some common types of folds with characteristic strike and dip patterns.

PART 1: STRIKE, DIP, CONTACTS AND GEOLOGIC MAPS

Geologic maps show three important pieces of geologic information:

◆ Formations: a continuous (or once continuous) layer of rock

◆ Contacts: the boundaries between rock units or formations

◆ Structures: physical arrangements of rock masses that often include deformational features like faults and folds

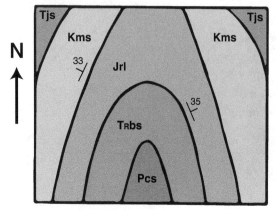

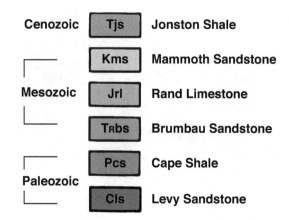

Figure 6. Geologic Map used for question 1.

What is Shown on a Geologic Map

1. Examine the geologic map and its key (figure   5).

    a. What type of rock does the symbol Jrl represent? _____

    b. What does the J in Jrl stand for? _____

    c. What about the rl? _____

    d. Name a formation that shares a contact with Jrl. _____

2. Examine the geologic map you used in question 1 and observe the strike and dip symbols. What type of geologic structure is shown in this map? Be specific.

3. Use the map (figure 7) to answer the following questions. For each lettered strike and dip symbol on the map, write out the approximate numeric strike and dip.

    a. _____   d. _____

    b. _____   e. _____

    c. _____

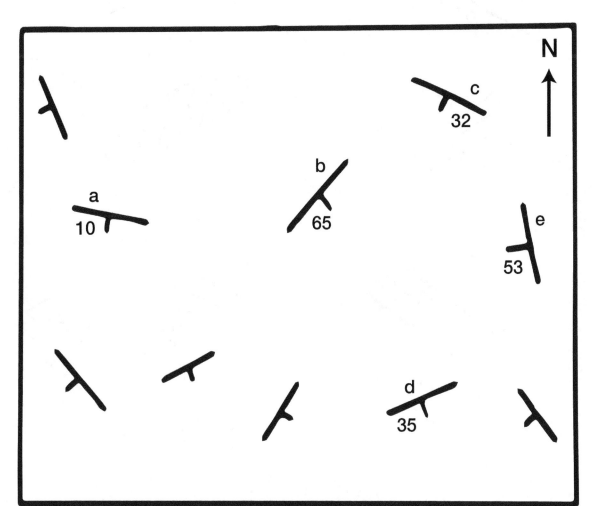

Figure 7. Map showing strike and dip symbols for question 3.

PART 2: BLOCK DIAGRAMS

Complete all blank sides of the following block diagrams using pencil and add appropriate strike and dip symbols. Check the Geologic Timescale (figure 7) for order of formation: S-Silurian, D-Devonian, M-Mississippian P-Permian.

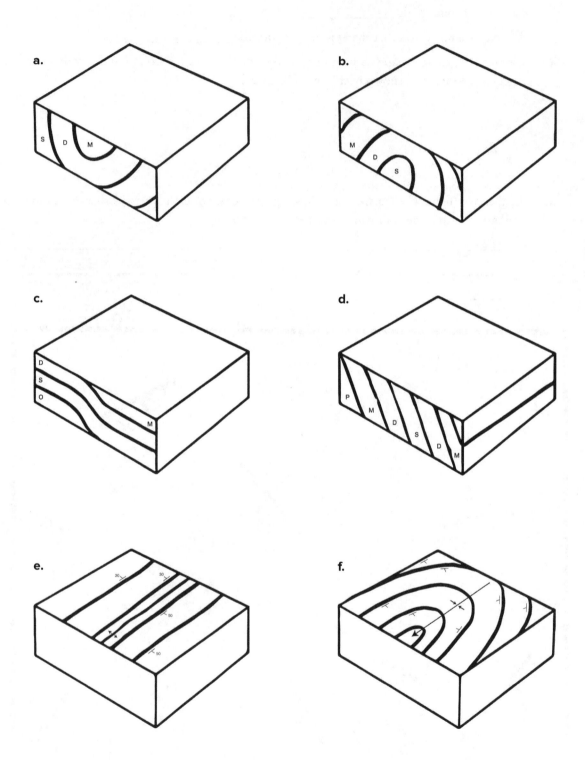

EON	ERA	PERIOD		EPOCH		Ma
Phanerozoic	Cenozoic	Quaternary		Holocene		
				Pleistocene	Late	0.011
					Early	0.8
		Tertiary	Neogene	Pliocene	Late	2.4
					Early	3.6
				Miocene	Late	5.3
					Middle	11.2
					Early	16.4
			Paleogene	Oligocene	Late	23.0
					Early	28.5
				Eocene	Late	34.0
					Middle	41.3
					Early	49.0
				Paleocene	Late	55.8
					Early	61.0
	Mesozoic	Cretaceous		Late		65.5
				Early		99.6
		Jurassic		Late		145
				Middle		161
				Early		176
		Triassic		Late		200
				Middle		228
				Early		245
	Paleozoic	Permian		Late		251
				Middle		260
				Early		271
		Pennsylvanian		Late		299
				Middle		306
				Early		311
		Mississippian		Late		318
				Middle		326
				Early		345
		Devonian		Late		359
				Middle		385
				Early		397
		Silurian		Late		416
				Early		419
		Ordovician		Late		423
				Middle		428
				Early		444
		Cambrian		Late		488
				Middle		501
				Early		513
Precambrian	Proterozoic	Late		Neoproterozoic (Z)		542
		Middle		Mesoproterozoic (Y)		1000
		Early		Paleoproterozoic (X)		1600
	Archean	Late				2500
		Early				3200
	Haydean					4000

Figure 8. Geologic timescale.

PART 3: FOLDS AND CROSS SECTIONS

Complete the cross sections using the maps and answer the questions. The higher the number, younger the rock.

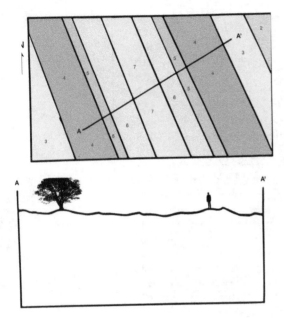

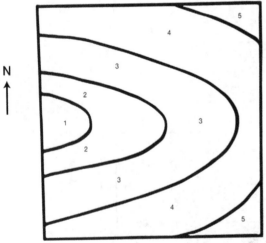

Figure 9. Map and cross section.

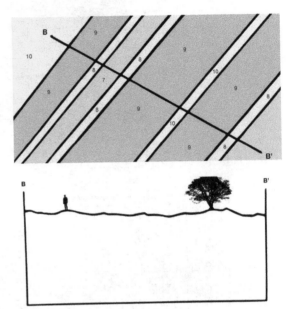

Figure 10. Map and cross section.

1. Is there an anticline or a syncline in your cross section? _____

2. Please include layers 5 and 6 in the cross sections in figure 9.

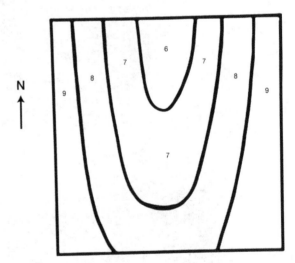

Figure 11.

Figure 12.

3. What type of fold is on this map?

_____

4. Direction of plunge for this fold?

_____

5. What type of fold is on this map?

_____

6. Direction of plunge for fold?

_____

PART 4: PUTTING IT ALL TOGETHER

Complete the cross section and answer the questions.

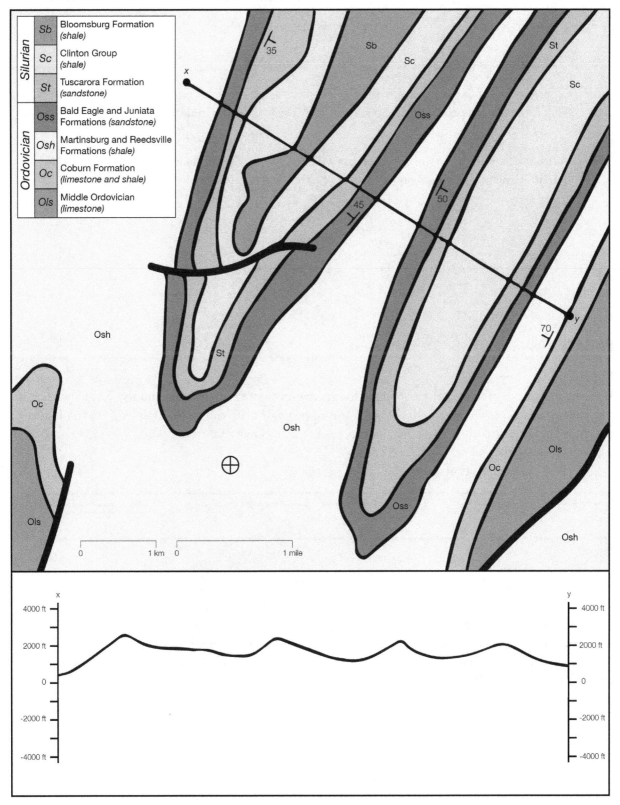

Figure 13. Geologic map and geologic cross section.

QUESTIONS

Answer these questions based on the geologic map in figure 11.

1. Based on the outcrop pattern formed by the Silurian rocks in the area, what kind of structure do these rock types form?

2. What is the general compass direction of strike for unit Oss? _____

3. What is the compass direction of dip for unit Oss where 50° dip is indicated? _____

4. Notice the fault in the approximate center of the map area. Did this fault form before or after the Silurian. Explain how you know.

PART 5: FAULTING EXERCISES

Forces

There are three types of force that act upon the Earth's crust: compression, tension, and shear. Compression is a squeezing force that shortens and thickens the crust. Tension is a stretching force that lengthens and thins the crust. Shear is a wrenching force that doesn't change the thickness of the crust but simply causes two adjacent blocks to slide laterally past each other.

We show the three types of forces with arrows as follows:

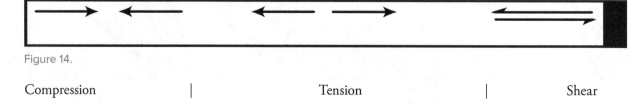

Figure 14.

Compression | Tension | Shear

The thick black stripe in each of the five diagrams on the next page represents a rock layer, illustrating the offset appropriate for each fault type. Fill in the fault name and indicate the dominant force involved in the space beside each diagram. Put directional arrows on the diagram to indicate the direction of offset. Then, label the Hanging Wall (HW) and the Foot Wall (FW).

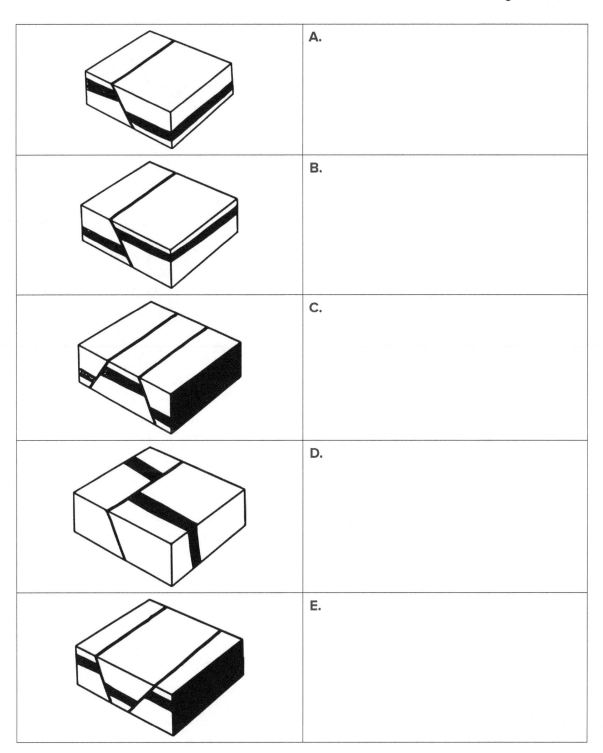

A.

B.

C.

D.

E.

Field Trip 1
142 Write-Up

Name _____   Date _____

STOP 1: CASCADE HEAD

◆   We are not going to hike to the top of this head, but if you get the chance, it's well worth it. However, from where we are parked, look out and slightly to the north. Draw the rock outcrop below. What type of rock do you think this is? Is it related to the Columbia River Basalts?

◆   From this new vantage point, describe and sketch the river we are looking out on. Is its gradient steep or flat? Is its velocity fast or slow? Relate the river's behavior back to our coastal tectonics and indicate whether the river is eroding vertically or laterally. How can its characteristics be changed?

STOP 2: PROPOSAL ROCK, NESKOWIN

◆ Get as close as you can to Proposal Rock. Describe the texture of the rock that you see. How is this rock different from other basaltic flows you've seen? What would have to happen to a basalt flow to create the rocks that you are seeing?

STOP 3: CAPE KIWANDA, PACIFIC CITY

◆ Compare the color of the cape to Haystack Rock in the ocean. Based on your observations, are they the same rock? What type of rock are the caps and Haystack Rock, respectively? Are they igneous, sedimentary, or metamorphic?

◆ How exactly do you think the sand dune formed behind the cape?

STOP 4: NESTUCCA VALLEY HIGH SCHOOL, CLOVERDALE

◆　Notice the topography of the surrounding valley. Describe it, making sure to mention the valley walls and the outlet for the Nestucca River. If there were a large magnitude earthquake and resulting tsunami, how would this valley's configuration affect the outward flow of tsunami waters?

STOP 5: MT, HEBO, HEBO (POSSIBLE STOP)

◆　Notice how flat Mt. Hebo is, what do you think has protected this particular peak from erosion? The highest peaks in the Oregon Coastal mountains have similar caps, allowing them to maintain their height, like Mary's Peak.

Field Trip 2
MSH Field Trip Report

Name _____   Date _____

KNOWLEDGE REVIEW

◆ Define **viscosity** and explain what influences the viscosity of magma.

◆ List the 3 aphanitic volcanic (extrusive) rocks we studied. Put them in order of increasing silica content.

◆ What is Dacite? Hint: it's not in your book – review your notes or ask an instructor after you have answered the second question.

DRIVING NORTH (I-5)

◆ Are any volcanoes visible from the commute? If so, name all that you view below.

◆ We cross into Washington state on the I-5 interstate bridge. What river are we crossing?

◆ What direction is the river flowing?

DRIVING EAST (STATE ROUTE 504)

◆ What types of rocks or lava flows do you see en route that we have identified in class?

◆ What is the name of the national forest we are entering as we approach JRO?

JOHNSTON RIDGE OBSERVATORY

◆ Who is the namesake of this observatory? Explain.

◆ Find the "Topographic model with colored lights." Listen to the description of the eruption and fill in the chart below. Indicate the colors of lights used and describe each event and how it changed the landscape:

List the Eruptive Events in Order	Color of Eruptive Event	Describe Event. How did it change MSH or the landscape?
1		
2		
3		
4		
5		
6		

◆ Find the exhibit "There are Different Recipes for Eruption." Read the panels and examine the rocks.

a. Circle the volcano that explodes more violently:
   Kilauea   OR   Mt. St. Helens

b. Which type of lava contains the least amount of silica?
   Basalt   OR   Dacite

◆ Find the panel "Lava Domes Rebuild Mt. St. Helens."

a. Circle which type of dacite lava is destructive:
   Gas-rich dacite lava   OR   Gas-poor dacite lava

b. Describe how the formation of the 1980–1986 and 2004–2008 lava domes were similar and different.

◆ Watch the video entitled "Crater Glacier."

   a. What feature created by the destructive May 18, 1980 eruption allowed the glacier to form?

   b. What do you think will happen to the crater glacier in the future – will it retreat or advance? Why?

◆ Go to the "Shattered Tree Stump" and find the display about tree rings. What do the tree rings tell us about MSH's eruptive history?

◆ Find the "New Answers to Old Mysteries" exhibit.

   a. What two eruptive events at MSH led to the discovery of similar events at other volcanoes?

   b. How is MSH rebuilding itself?

ERUPTION TRAIL

Stop 1: Find a stump or log beside the trail between the trailhead and the first interpretive sign. Answer these quesions.

◆ Towering evergreen trees grew here before the eruption. The lateral blast created 3 distinct "impact zones." Make observations to determine the impact zone here. Circle one.

   a. Tree Removal Zone: Trees uprooted, shattered, and swept away from 0-7.5 miles away from the volcano.

   b. Blowdown Zone: Trees are blown down in distinct patterns. Ridges and topography direct the blast 2.5 to 15.5 miles away from the volcano.

   c. Scorch Zone: Hot gasses scorch trees, creating a standing dead forest, 2.5 to 17 miles.

Stop 2: A sundial-like landscape locator is at the circular plaza on the top of the hillside.

◆ Find Coldwater Peak on the landscape locator and observe the blast features below and around it. Circle which "impact zone" in which Coldwater Creek belongs.

   a. Tree Removal Zone   b. Blowdown Zone   c. Scorch Zone

- Explain how a blown-down tree is like a compass needle.

- Look around for blown down trees on the nearby ridges. Use the locator map to identify what direction the blown trees lay (point towards). _____

- Use the locator map and find east. Look east down the length of Johnston Ridge and find the mounds on top of the ridge. What are these called and why are they there?

HUMMOCKS TRAIL

Stop 1: After reading the background information below, look at the trailside rocks by the bench. Based on the color of rocks in your surroundings, from which slide block did these rocks originate? Circle one.

    a.   Slide Block I   b.   Slide Block II   c.   Slide Block III

Stop 2: Johnston Ridge should be in sight and MSH is straight ahead. The rocks surrounding the trail came from the same slide block as above.

- The landslide began on the north side of the volcano, but traveled only 5.5 miles north. However, it traveled 13 miles westward. Why do you think this happened?

Stop 3: The two-toned hummock.

- Notice that the hummock that is half light tan and half blackish-red. Based on the color, name the types of rocks that might be found on each side of this hummock.

SOME BACKGROUND INFORMATION

The May 18, 1980 eruption began when a huge landslide fell from the bulging north side of Mount St. Helens. Three enormous slabs of rock, called slide blocks, fell downward in quick succession.

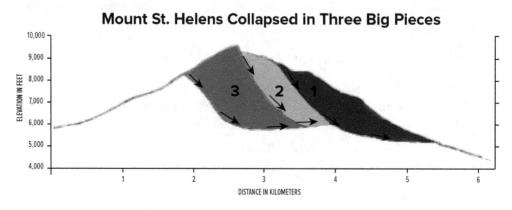

Mount St. Helens Collapsed in Three Big Pieces

Slide Block I is dominated by dark gray andesite lava rocks and black or red basalt lava rocks from the north flank of the volcano. It also contains minor amounts of light gray dacite rocks from the summit.

Slide Block II is dominated by light gray dacite rocks from the former summit.

Moments after Slide Block I came to rest, Slide Block II pushed it out of the way like a gigantic snow plow. Due to this violent collision, slide block I is largely found along the sides of the valley, while side block II dominates the center of the valley.

Slide Block III is dominated by tan dacite lava rocks from the interior of the volcano. This slide block was violently shoved outward by a lateral blast, pushing the landslide 13½ miles down valley (4 miles past slide blocks I and II). Few intact hummocks from this slide block are found west of the hummocks trail—they were broken apart and mixed together during their turbulent journey.

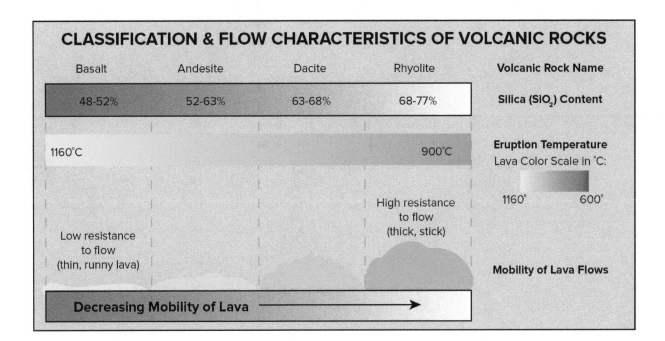

Acknowledgments

TEXT ACKNOWLEDGMENTS

Lab 13

"Geologic Structures" by Michelle Harris is copyrighted work and used by permission in this work only. It may not be used outside of this work or in any derivative work without further permission.

IMAGE ACKNOWLEDGMENTS

These chapters incorporate images from the following sources.

Lab 1

Figure 1.1. "Belknap Crater Area" by Shannon Othus-Gault is a product of work by Chemeketa Press.

Figure 1.2. "Smith Rock rhyolite and Newberry Basalt" by Shannon Othus-Gault is a product of work by Chemeketa Press.

Figure 1.3a. "Aerial Crater Lake" by Mike Doukas is licensed under Public Domain (https://commons.wikimedia.org/wiki/File:Aerial_Crater_Lake.jpg).

Figure 1.3b. "Wizard Island, Crater Lake" by Shannon Othus-Gault is a product of work by Chemeketa Press.

Figure 1.4. "Yellowstone Panorama" by Reanna Camp is a product of work by Chemeketa Press.

Figure 1.5. "The Spires, Crater Lake" by Shannon Othus-Gault is a product of work by Chemeketa Press.

Lab 2

Figure 2.1. "Ocean Floor" by NOAA is licensed under Public Domain (http://oceanexplorer.noaa.gov/edu/learning/2_midocean_ridges/activities/seafloor_spreading.html).

Figure 2.2. "Magnetic Anomolies" by NOAA is licensed under Public Domain (http://oceanexplorer.noaa.gov/edu/learning/2_mid-ocean_ridges/activities/seafloor_spreading.html).

Figure 2.3. "Political Map of Medditerranean Sea" by Nations Online Project is licensed under Free for Educational use (http://www.nationsonline.org/oneworld/map/Mediterranean-Region-Map.htm).

Figure 2.4. "Age of Hawaiian Islands, by Cierra Maher is a product of work by Chemeketa Press.

Figure 2.5. "Juan de Fuca Plate Motion" by Cierra Maher is a product of work by Chemeketa Press (http://commons.bcit.ca/civil/students/earthquakes/unit1_02.htm).

Figure 2.6. "Volcanic Activity map" by Dale Sawyer is licensed under Copyright (http://plateboundary.rice.edu/downloads.html).

Figure 2.7. "Earthquake Depth and Distribution map" by Dale Sawyer is licensed under Copyright (http://plateboundary.rice.edu/downloads.html).

Lab 3

Figure 3.1. "Determining hardness with a glass plate" by Ronald Cox IV is a product of work by Chemeketa Press.

Figure 3.2. "Some of the common shapes that minerals break into along cleavage planes" by Reanna Camp is a product of work by Chemeketa Press.

Figure 3.3. "A few of the many color varieties of quartz" by Ronald Cox IV is a product of work by Chemeketa Press.

Figure 3.4. "Galena - Streak color" by Ra'ike is licensed under GNU, CC BY-SA 3.0, 2.5, 2.0, 1.0 (https://commons.wikimedia.org/w/index.php?curid=17279903).

Figure 3.5. "The mineral calcite vigorously reacts with acid" by Ronald Cox IV is a product of work by Chemeketa Press.

Figure 3.6. "Calcite showing double refraction." by Ronald Cox IV is a product of work by Chemeketa Press.

Lab 5

Figure 5.1. "Some common igneous rocks and mineral compositions" by Cierra Maher is a product of work by Chemeketa Press.

Lab 10

Figure 10.1. "Major Cascade Range volcanoes. Mt. St. Helens is shown in red" by USGS is licensed under Public Domain (https://pubs.usgs.gov/gip/msh/cascade.html).

Figure 10.2. "Digital map of MSH before (above) and after (below) the 1980 eruption" by USGS is licensed under Public Domain (https://pubs.usgs.gov/circ/c1050/century.htm).

Figure 10.3. "Making a transect line" by Cierra Maher is a product of work by Chemeketa Press.

Figure 10.4. "Creating a profile on a graph" by Cierra Maher is a product of work by Chemeketa Press.

Figure 10.5. "Topographic map of Mt. St. Helens before 1980 eruption" by USGS is licensed under Public Domain (https://volcanoes.usgs.gov/volcanoes/st_helens/st_helens_gallery_27.html).

Figure 10.6. "MSH pre-eruption profile" by Cierra Maher is a product of work by Chemeketa Press.

Figure 10.7"Topographic map of Mt. St. Helens after 1980 eruption" by USGS is licensed under Public Domain (https://volcanoes.usgs.gov/volcanoes/st_helens/st_helens_gallery_27.html).

Figure 10.8. "MSH post-eruption profile" by Cierra Maher is a product of work by Chemeketa Press.

10.9. "MSH pre- and post-eruption profiles" by Cierra Maher is a product of work by Chemeketa Press.

Lab 11

Figure 11.1. "Three Seisomograms" by Cierra Maher is a product of work by Chemeketa Press.

Figure 11.2. "S-P graph" by Cierra Maher is a product of work by Chemeketa Press.

Figure 11.4. "Western US map" by Cierra Maher is a product of work by Chemeketa Press.

Figure 11.5. "Nomogram" by Cierra Maher is a product of work by Chemeketa Press.

Figure 11.5. "Earthquake Depth and Distribution map" by Dale Sawyer is licensed under Copyright (http://plateboundary.rice.edu/downloads.html).

Figure 11.6. "Volcanic Activity map" by Dale Sawyer is licensed under Copyright (http://plateboundary.rice.edu/downloads.html).

Figure 11.7. "Wadati-benioff-zone" by RyanMikulovsky is licensed under Public Domain (https://en.wikipedia.org/wiki/File:Wadati-benioff-zone.png).

Figure 11.8. "Tonga Trench" by NOAA is licensed under Public Domain (http://noaa.gov).

Figure 11.9. "Tonga Graph" by Reanna Camp is a product of work by Chemeketa Press.

Lab 12

Figure 12.1. "Laurasia-Godwana" by LennyWikipedia is licensed under CC-BY 3.0 (https://commons.wikimedia.org/wiki/File:Laurasia-Gondwana.svg).

Figure 12.2. "Geologic time scale" by USGS is licensed under Public Domain (https://commons.wikimedia.org/wiki/File:Geologic_time_scale.jpg).

Figure 12.3. "Accreted Terrane Map" by Shannon Othus-Gault and Cierra Maher is a product of work by Chemeketa Press.

Figure 12.4. "Accreted Terranes" by Shannon Othus-Gault and Cierra Maher is a product of work by Chemeketa Press.

Lab 13

Unless otherwise noted, images in this lab were created by Ronald Cox IV and are a product of work by Chemeketa Press.

Figure 13.7. "Geologic Timescale" by USGS is licensed under Public Domain (https://commons.wikimedia.org/wiki/File:Geologic_time_scale.jpg).

Field Trip 2

Figure FT2.1. "MSH collapse" by Cierra Maher is a product of work by Chemeketa Press.

Figure FT2.2. "Classification and Flow Characteristics of volcanic rocks" by Cierra Maher is a product of work by Chemeketa Press.

CPSIA information can be obtained
at www.ICGtesting.com
Printed in the USA
LVHW022350201119
637811LV00002B/16/P